CHARACTERIZATIONS OF EXPONENTIAL DISTRIBUTION BY ORDERED RANDOM VARIABLES

APPLIED STATISTICAL SCIENCE

MOHAMMAD AHSANULLAH - SERIES EDITOR

PROFESSOR, RIDER UNIVERSITY, LAWRENCEVILLE, NEW JERSEY, UNITED STATES

Characterizations of Exponential Distribution by Ordered Random Variables
Mohammad Ahsanullah (Editor)
2019. ISBN: 978-1-53615-402-3
(Softcover)

Applied Statistical Theory and Applications
Mohammad Ahsanullah (Editor)
2014. ISBN: 978-1-63321-858-1
(Hardcover)

Research in Applied Statistical Science
Mohammad Ahsanullah (Editor)
2014. ISBN: 978-1-63221-818-5
(Hardcover)

The Future of Post-Human Probability Towards a New Theory of Objectivity and Subjectivity
Peter Baofu (Author)
2014. ISBN: 978-1-62948-671-0
(Hardcover)

Sequencing and Scheduling with Inaccurate Data
Yuri N. Sotskov and Frank Werner (Editors)
2014. ISBN: 978-1-62948-677-2
(Hardcover)

Dependability Assurance of Real-Time Embedded Control Systems
Francesco Flammini (Author)
2010. ISBN: 978-1-61728-502-8
(Softcover)

APPLIED STATISTICAL SCIENCE

CHARACTERIZATIONS OF EXPONENTIAL DISTRIBUTION BY ORDERED RANDOM VARIABLES

MOHAMMAD AHSANULLAH
RIDER UNIVERSITY

nova
science publishers
New York

NOTICE TO THE READER

Library of Congress Cataloging-in-Publication Data

Names: Ahsanullah, Mohammad, author.
Title: Characterizations of exponential distribution by ordered random variables / Mohammad Ahsanullah (Rider University, Lawrenceville, NJ, USA).
Description: Hauppauge, New York : Nova Science Publishers, Inc., 2019. |
Series: Applied statistical science | Includes bibliographical references.
Identifiers: LCCN 2019017069 (print) | LCCN 2019022056 (ebook) | ISBN 9781536154030 (ebook) | ISBN 9781536154023 (softcover)
Subjects: LCSH: Distribution (Probability theory) | Variables (Mathematics)
Classification: LCC QA273.6 (ebook) | LCC QA273.6 .A4337 2019 (print) | DDC 519.2/4--dc23
LC record available at https://lccn.loc.gov/2019017069

Published by Nova Science Publishers, Inc. † New York

To Masuda, Angela, Annie, Nisar, Tabassum,
Omar Faruk. Zak, Sam, Amil and Julian

CONTENTS

PREFACE

Exponential distribution is widely used in applied Science. The exponential distribution is often used to model failure times of manufactured items in production. It is the only continuous distribution that has memoryless property.

We may define the memoryless property of a random variable as the future event is independent of the past event. Many times using data we like to know its distribution. There are various methods to characterize the distribution using the data. In this book many different methods are given to characterize the exponential distribution.

Chapter 1 gives some of the basic properties of the exponential distribution. Chapter 2 gives the characterizations of the exponential distribution by order statistics. Chapter 3 presents the characterizations of the exponential distributions by record values and chapter 4 gives the characterizations of the exponential distribution by generalized order statistics. Extensive references dealing with characterizations of exponential distributions by order random variables are given.

A Summer research grant and sabbatical leave from Rider University enabled me to complete the book.

I am forever grateful to my wife for her continuous encouragement and unwavering support through the years.

Chapter 1

BASIC PROPERTIES

1.1. INTRODUCTION

The exponential distribution (also known as negative exponential distribution) is one of the most important univariate continuous distributions having various applications. Exponential distribution is the only continuous distribution that has memoryless property. A random variable is said to have memoryless property if $P(X > s + t | X > t) = P(X > s)$ for $0 < s, t < \infty$. A random variable X is said to have an exponential distribution with location parameter μ and scale parameter σ if the probability density function (pdf) $f_{\mu,\sigma}(x)$ of X is as follows:

$$f_{\mu,\sigma}(x) = \sigma e^{-\sigma(x-\mu)}, x \leq \mu, \sigma > 0,$$
$$= 0, \text{ otherwise.} \tag{1.1.1}$$

We say that X is distributed as $E(\mu, \sigma)$ if the pdf is as given in (1.1). The pdf of X is monotonically decreasing from σ to 0 as X increases from μ to ∞. The mean of X is $\mu + 1/\sigma$, the mode is σ and it occurs at $x = \mu$. If $\mu = 0$ and $\sigma = 1$ then we say that X has the standard exponential distribution. The pdf of standard exponential distribution is as follows:

$$f_{0,1}(x) = e^{-x}, x \geq 0$$
$$= 0, \text{ otherwise.} \tag{1.1.2}$$

The following Figure 1.1.1. gives the pdf of X for $\mu = 0$ and $\sigma = 0.5$, 1.0 and 2.0.

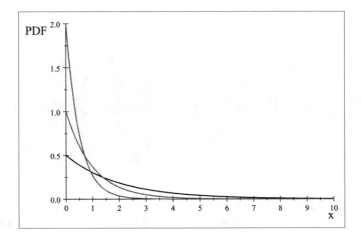

Figure 1.1.1. PDFs of E(0, σ), Black for $\sigma = 0.5$, Red for $\sigma = 1.0$ and Green for $\sigma = 2.0$.

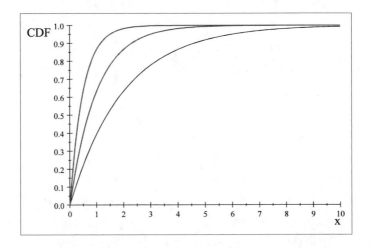

Fig. 1.1.2. CDFs of E(0, σ), Black for $\sigma = 0.5$, *Red for* $\sigma = 1$ and Green for $\sigma = 2$.

The cumulative distribution function (cdf) $F_{\mu,\sigma}(x)$ is given as follows:

$$F_{\mu,\sigma}(x) = 0 \text{ for } X < \mu$$
$$= 1 - e^{-\sigma(x-\mu)} \text{ for } x \geq \mu.$$

The Figure 1.1.2. gives the cdf of X for $\mu = 0$ and $\sigma = 0.5, 1.0$ and 2.0.

The hazard rate r(x) of a random variable is defined as $\frac{f(x)}{1-F(x)}$ for $0 < F(x) < 1$. The hazard rate of the exponential distribution with pdf as given in (1. I) is σ. Let $\mu_r = E(X^r)$, then

$$\mu_r = \int_\mu^\infty x^r \sigma e^{-\sigma(x-\mu)} dx$$
$$= \int_0^\infty (\mu + \frac{u}{\sigma})^r \, e^{-u} \, du$$
$$= \sum_{k=0}^r \frac{r!}{k!(r-k)!} \frac{1}{\sigma^k} \mu^{r-k} \Gamma(k+1)$$

We have

$$E(X) = \mu + \frac{1}{\sigma},$$
$$E(X^2) = \mu^2 + \frac{2\mu}{\sigma} + \frac{2}{\sigma^2},$$

and

$$Var(X) = \frac{1}{\sigma^2}.$$

We define the memoryless property of a random variable X's by the relation, for all s, t > 0.

$$P(X > s + t | X > t) = P(X > s). \tag{1.1.3}$$

It can easily be seen that for the exponential distribution P(X > s + t|X > t) = P(X > s). This is equivalent to (1-F(s + t)) = (1-F(s))(1-F(t)). This is called the memoryless property. The exponential distribution is the only continuous distribution that has this property. We have the following characterization theorem based on the memoryless property.

Theorem 1.1.1.

Let X be a non-negative continuous random variable having cdf F(x) with F(0) = 0. If X satisfies the condition given in (1.1.3), then F(x) = $1 - e^{-\lambda x}$, for all x $\geq$ 0 and any λ > 0.

Proof

Since P(x = 0})≠1, then P(X > t) > 0 for some t > 0 and we can write the equation (1.1.3) as

$$P(X \geq s + t) = P(X \geq s)P(X \geq t) \text{ for all s,t} \geq o \tag{1.1.4}$$

i.e.:

$$1 - F(s + t) = (1-F(s))(1-F(t)). \text{ for all s,t} \geq 0. \tag{1.1.5}$$

Writing G(x) = 1-F(x), we obtain x > 0

$$G(s + t) = G(s) G(t), \text{ for all s,t} > 0. \tag{1.1.6}$$

This is the well-known Cauchy functional equation. The continuous solution of (1.1.6) (see Aczel [1966]) is

$$G(x) = e^{bx}, \text{ for all all x} \geq 0 \text{ and any real b.} \tag{1.1.7}$$

Since F(x) is a cdf with $F(0) = 0$ and $F(\infty) = 1$, we must have

$F(x) = 1 - e^{-\lambda x}$, for all x $\underline{10}$ 0 and any real $\lambda > 0$.   (1.1.8)

The following is a characteristic property of the exponential distribution.

For an absolutely continuous random variable having cdf F(x) with $F(0) = 0$, pdf f(x) and the condition $Var(X|X > y) = c$, for all $y > 0$, where c is a constant characterizes the exponential distribution. We can write $Var(X|X > y) = c$ as

$E\,(X - y)^2|X > y) = c$ for all $y \geq 0$.   (1.1.9)

However, the following relation (see Dallas [1979]) characterizes the exponential distribution.

$E(X - y)^r|X > y) = c, r \geq 1$, for all $y > 0$.   (1.1.10)

The equation (1.1.10) will lead to the equation

$\int_y^\infty r(x - y)^{r-1} (1 - F(x))dx = c(1 - F(y))$ (1.1.11)

Dallas [1979] showed that the solution of equation (1.1.11) is

$F(x) = 1 - \exp(-\lambda x,)$, for all $x \geq 0$ and any $\lambda > 0$.

For the exponential distribution with $f(x) = e^{-x}, x \geq 0$, $E(X|X \leq X\ )$ $= 1 + \frac{x}{1 - e^x}$. This condition is a characteristic property of the exponential distribution. Assume X is a positive continuous random variable with $F(0) = 0$ and $F(X) > 0$ for all $x > 0$. Suppose we have $E(X|X \leq x\ ) = 1 +$

$\frac{x}{1-e^x}$, then writing $\int_0^x uf(u)du = F(x)\,(1 + \frac{x}{1-e^x})$ and differentiating the both sides of the equation, we obtain on simplification,

$$\frac{f(x)}{F(x)} = \frac{e^{-x}}{1-e^{-x}} \tag{1.1.12}$$

On integrating both sides of the equation and using the condition F(0) = 0 and F(∞) = 1, we obtain

F(x) = 1-e^{-x}, x ≥ 0.

Suppose the random variable X has an absolutely continuous distribution with cdf F(x) and pdf f(x). Let F(0) = 0 and s(x) = 1-F(x). Grosswald and Kotz [1979] showed that if

$$\int_0^\infty f(x)(\frac{S(x+z)}{S(z)} - s(z))\,dx = 0$$

Then F(x) = 1-$e^{-\lambda x}$, for some $\lambda > 0$ and x≥ 0.
Huang and Su [2013] proved the following Theorem.

Theorem 1.1.2.

Suppose X is an absolutely continuous random variable with cdf F(x) and pdf f(x).If F(0) = 0 and F(x) > 0 for all x > 0, then

$$E((X-x)^2|X > x)) - 2\,(E(X-x)|X > x)^2 = 0$$

If and only if F(x) = 1-$e^{-\lambda x}$, $\lambda > 0$, x ≥ 0.

Suppose the random variable X has exponentiated exponential distribution (EEXP) with the pdf as

$$f_{EEXP}(x) = \alpha\beta e^{-\beta x}\left(1 - e^{-\beta x}\right)^{\alpha-1}, x \geq 0, \alpha, \beta > 0. \tag{1.1.13}$$

If $\alpha = 1$, then *EEXP is EXP*.

The following figure 1.1.4. gives cdf of EEXP for $\beta = 1$ *and* $\alpha = 2, 4$ and 6.

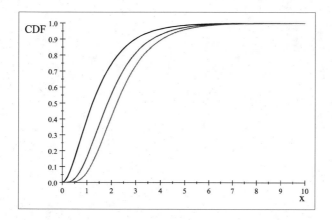

Figure 1.1.3. CDF of X $_{EEXP}$.for β=1. Black for α=2, red for α=4 and green for α=6.

Saran and Pandey [2004] gave some recurrence relations of the moments of X_{EEXP}..

The following theorem is based on conditional right truncated first moment of the random variable X.

Theorem 1.1.3.

Suppose that the random variable X is continuous (with respect to Lebesgue measure) with cdf F(x) and pdf f(x) and F(0)= 0 and $F(\infty) = 1$.

We assume E(x) is finite. Then $EX|X \leq x) = g(x)\tau(x)$ where $\tau(x) = \dfrac{f(x)}{F(x)}$

and

$$g(x) = \sqrt{\frac{\pi}{2}} \frac{e^{\frac{1}{2}(x+1)^2}}{1+x} \left(\text{erf}\left(\frac{x+1}{\sqrt{2}}\right) - \text{erf}\left(\frac{1}{\sqrt{2}}\right) \right) - \frac{x}{1+x},$$

if and only if

$$f_{le}(x) = (1+x)e^{-(x+\frac{x^2}{2})}, x \geq 0.$$

Proof of Theorem 1.1.3.

If $f_{le}(x) = (1+x)e^{-(x+\frac{x^2}{2})}$, then

$$f_{le}(x)g(x) = \int_0^x u(1+u)e^{(u+\frac{u^2}{2})}du$$

$$= \sqrt{\frac{\pi}{2}} e^{\frac{1}{2}} \left(\text{erf}\left(\frac{x+1}{\sqrt{2}}\right) - \text{erf}\left(\frac{1}{\sqrt{2}}\right) \right) - x\, e^{-(x+\frac{x^2}{2})}$$

Thus

$$g(x) = \sqrt{\frac{\pi}{2}} \frac{e^{\frac{1}{2}(x+1)^2}}{1+x} \left(\text{erf}\left(\frac{x+1}{\sqrt{2}}\right) - \text{erf}\left(\frac{1}{\sqrt{2}}\right) \right) - \frac{x}{1+x}$$

Suppose

$$g(x) = \sqrt{\frac{\pi}{2}} \frac{e^{\frac{1}{2}(x+1)^2}}{1+x} \left(\text{erf}\left(\frac{x+1}{\sqrt{2}}\right) - \text{erf}\left(\frac{1}{\sqrt{2}}\right) \right) - \frac{x}{1+x}, \text{then}$$

$$g'(x) = x + \left(\sqrt{\frac{\pi}{2}} \frac{e^{\frac{1}{2}(x+1)^2}}{1+x} \left(\text{erf}\left(\frac{x+1}{\sqrt{2}}\right) - \text{erf}\left(\frac{1}{\sqrt{2}}\right) \right) - \frac{x}{1+x}\right)(\frac{x(x+2)}{x+1}).$$

$$= x + g(x)\,(\frac{x(x+2)}{x+1})$$

Thus

$$\frac{x - g'(x)}{g(x)} = -\frac{x(x+2)}{x+1}.$$

By Lemma A.1, as given in the Appendix, we have

$$\frac{f'(x)}{f(x)} = -\frac{x(x+2)}{x+1}$$

On integrating both sides of the above equation with respect to x, we obtain

$$f(x) = c(1+x)e^{-(x+\frac{x^2}{2})},$$

where c is a constant.

Using the boundary condition $F(0) = 0$ and $F(\infty) = 1, we$ have

$$f(x) = (1+x)e^{-(x+\frac{x^2}{2})}, x \geq 0.$$

The following theorem is based on the left truncated first moment.

Theorem 1.1.4.

Suppose the random variable X has continuous (with respect to Lebesgue measure) cdf F(x), pdf f(x), $F(0) = 0$ and $F(\infty) = 1$,

We assume E(x) is finite. Then $E(X|X > x) = h(x)r(x))$ where $r(x) = \frac{f(x)}{1-F(x)}$, $h(x) = \frac{E(x)}{f(x)} - g(x)$ and g(x) is as given in Theorem 1.1.2.

If and only if

$$f(x) = (1+x)e^{-(x+\frac{x^2}{2})},$$

Proof of Theorem 1.1.4.

$$f_{le}(x) = (1 + x)e^{-(x+\frac{x^2}{2})}, then$$

$$f_{le}(x)h(x) = \int_x^\infty u(1 + u)\, e^{-(u+\frac{u^2}{2})}du$$

$$E(X) - \int_0^x u(1 + u)\, e^{-(u+\frac{u^2}{2})}du$$

$$= \sqrt{(\tfrac{\pi}{2})}e^{\frac{1}{2}}(1 - \operatorname{erf}(\tfrac{1}{\sqrt{2}}))$$

$$- \sqrt{\tfrac{\pi}{2}}\, e^{\frac{1}{2}}\left(\operatorname{erf}\left(\tfrac{x+1}{\sqrt{2}}\right) - \operatorname{erf}\left(\tfrac{1}{\sqrt{2}}\right)\right) - x\, e^{-(x+\frac{x^2}{2})}$$

Thus

$$h(x) = \sqrt{(\tfrac{\pi}{2})}e^{\frac{1}{2}}(1 - \operatorname{erf}(\tfrac{1}{\sqrt{2}}))\, \frac{e^{(x+\frac{x^2}{2}}}{1+x)}$$

$$- \sqrt{\tfrac{\pi}{2}}\, \frac{e^{\frac{1}{2}(x+1)^2}}{1+x}\left(\operatorname{erf}\left(\tfrac{x+1}{\sqrt{2}}\right) - \operatorname{erf}\left(\tfrac{1}{\sqrt{2}}\right)\right) - \frac{x}{1+x}$$

Suppose

$$h(x) = (\sqrt{(\tfrac{\pi}{2})}e^{\frac{1}{2}}(1 - \operatorname{erf}(\tfrac{1}{\sqrt{2}}))\, \frac{e^{(x+\frac{x^2}{2}}}{1+x)}$$

$$- \sqrt{\tfrac{\pi}{2}}\, \frac{e^{\frac{1}{2}(x+1)^2}}{1+x}\left(\operatorname{erf}\left(\tfrac{x+1}{\sqrt{2}}\right) - \operatorname{erf}\left(\tfrac{1}{\sqrt{2}}\right)\right) - \frac{x}{1+x}, then$$

$$h'(x) = -x + ((\sqrt{(\tfrac{\pi}{2})}e^{\frac{1}{2}}(1 - \operatorname{erf}(\tfrac{1}{\sqrt{2}}))\, \frac{e^{(x+\frac{x^2}{2}}}{1+x)}$$

$$- \sqrt{\tfrac{\pi}{2}}\, \frac{e^{\frac{1}{2}(x+1)^2}}{1+x}\left(\operatorname{erf}\left(\tfrac{x+1}{\sqrt{2}}\right) - \operatorname{erf}\left(\tfrac{1}{\sqrt{2}}\right)\right) - \frac{x}{1+x}\, \frac{x(x+2)}{x+1}$$

$$= -x + h(x)\left(\frac{x(x+2)}{x+1}\right)$$

Thus

$$\frac{x + h\prime(x)}{h(x)} = \frac{x(x + 2)}{x + 1}$$

and by Lemma A.2 as given in the Appendix, we have

$$\frac{f\prime(x)}{f(x)} = -\frac{x(x + 2)}{x + 1}$$

On integrating both sides of the above equation, we obtain

$$f(x) = c(1 + x)e^{-(x + \frac{x^2}{2})},$$

where c is a constant.

Using the boundary condition $F(0) = 0$ and $F(\infty) = 1,$ *we* have

$$f(x) = (1 + x)e^{-(x + \frac{x^2}{2})}, \; x \geq 0.$$

Marshall and Olkin (1977) introduced a powerful method of adding a new parameter to an existing distribution. If we have a family of distribution with cdf H(x), then the Marshall -Olkin extended cdf F(x) is as follows.

$$F(x) = \frac{H(x)}{\alpha + (1-\alpha)H(x)}, \; -\infty < x < \infty, \alpha > 0.$$

The cdf $F_{moe}(X)$ of the Marshall -Olkin exponential distribution (MOE) will be

$$F_{moe}(X) = 1 - \frac{\alpha e^{-x}}{1 - (1-\alpha)e^{-x}}, \; 0 \leq x < \infty, \alpha > 0. \tag{1.1.15}$$

The corresponding pdf $f_{MOE}(x)$ is given as follows:

$$f_{MOE}(x) = \frac{\alpha e^{-x}}{(1-(1-\alpha)e^{-x})^2}, 0 \leq x < \infty, \alpha > 0. \tag{1.1.16}$$

When $\alpha = 1, then$ MOE reduces to the standard exponential distribution and when $\alpha = 2$, then MOE reduces to half logistic distribution.

If the random variable X has the pdf as given in (1.1.16), then

$E(X) = \frac{\alpha ln\alpha}{\alpha - 1}, \alpha > 0, \alpha \neq 1.$

$E(X^k) = \Gamma(k + 1),$ if $\alpha = 1$

$= \alpha \Gamma(k + 1) \sum_{j=0}^{\infty} \frac{(1-\alpha)^j}{(j + 1)^k}, 0 < \alpha < 2. \alpha \neq 1.$

The median = $\ln(1 + \alpha), \alpha > 0.$

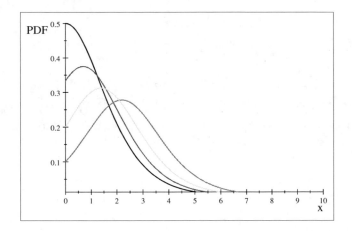

Figure 1.1.4. PDF $f_{MOE}(x)$: *Black for* $\alpha = 2, Red \, for \, \alpha = 3, Green \, for \, \alpha = 5$ and *Brown for* $\alpha = 10.$

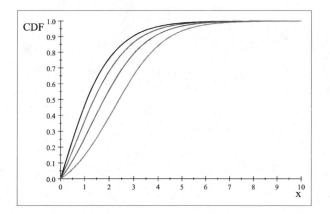

Figure 1.1.5. CDF FMOE(x), $Black\ for\ \alpha\ =\ 2, Red\ for\ \alpha\ =\ 3, Green$.

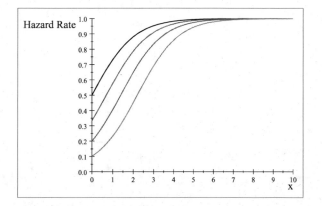

Figure 1.1.6. r(x), black for $\alpha\ =\ 0.2$, red for $\alpha = 0.6$,green for $\alpha = 2$ and brown for $\alpha = 6$.

Theorem 1.1.5.

Let $\{X_i, i \geq 1\}$ be a sequence of independent and identically distributed random variables with common survival function $G(x) = e^{-x}, x > 0$ and N be a geometric random variable with parameter α and $P(N = n) = \alpha(1 - \alpha)^{n-1}, n = 1,2, ...$ which is independent of $\{X_i\}$, for all i>1. Let $U_n = \min(X_i)$. Then $\{U_n\}$ is distributed as MOE.

Proof

Let $H(x) = P\{U_n \geq x\}$

$$= \sum_{k=1}^{\infty} \alpha(1-\alpha)^{n-1} e^{-x}$$

$$= \frac{\alpha}{e^x - (1-\alpha)}$$

which is MOE.

The Table 1.1. gives the percentile points x_p of MOE for $\alpha = 1,2,5$ and 10.

Table 1.1. Percentile points of MOE

	$\alpha = 1$	$\alpha = 2$	$\alpha = 5$	$\alpha = 10$
p	x_p	x_p	x_p	x_p
0.1	0.1054	0.2007	0.2877	0.7462
0.2	0.2231	0.4055	0.5596	1.2528
0.3	0.3567	0.6190	0.8267	1.6650
0.4	0.5108	0.8473	1.4663	2.0369
0.5	0.6912	1.0986	1.7918	2.3979
0.6	0.9163	1.3863	2.1401	2.7726
0.7	1.2040	1.7346	2.5390	3.1918
0.8	1.6094	2.1972	3.0445	3.7136
0.9	2.3036	2.9444	3.8786	3.5109

From the table, it is seen that as α increases the percentile points increase.

The following theorem gives a characterization of MOE based on right truncated first moment.

Theorem 1.1.6.

Suppose the random variable X has an absolutely continuous (with respect to Lebesgue measure) with cdf F(x), pdf f(x), F(0) = 0 and

$F(\infty) = 1$. We assume $E(x)$ is finite. Then $E(X|X \lesssim x) = g(x)\tau(x)$

where $\tau(x) = \dfrac{f(x)}{F(x)}$ and

$$g(x) = -x(1 - (1 - \alpha)e^{-x})$$
$$+ \frac{(1-(1-\alpha)e^{-x})^2}{\alpha e^{-x}} \frac{\alpha}{1-\alpha}(\ln(1 - (1 - \alpha)e^{-u}) - \ln\alpha)$$

if and only if

$$f(x) = \frac{\alpha e^{-x}}{(1-(1-a)e^{-x})^2}, \alpha > 0, 0 < x < \infty.$$

Proof

Suppose

$$f(x) = \frac{\alpha e^{-x}}{(1-(1-\alpha)e^{-x})^2}, \text{ then}$$

$$g(x) = \frac{\int_0^x \frac{\alpha u e^{-u}}{(1-(1-\alpha)e^{-u})^2}}{\frac{\alpha e^{-x}}{(1-(1-\alpha)e^{-x})^2}}$$

$$= -x(1 - (1 - \alpha)e^{-x}) + \frac{(1-(1-\alpha)e^{-x})^2}{\alpha e^{-x}} \int_0^x \frac{\alpha e^{-u}}{1-(1-\alpha)e^{-u}} du$$

$$= -x(1 - (1 - \alpha)e^{-x})$$

$$+ \frac{(1-(1-\alpha)e^{-x})^2}{\alpha e^{-x}} \frac{\alpha}{1-\alpha}(\ln(1 - (1 - \alpha)e^{-x}) - \ln\alpha).$$

$$g'(x) = -x(1 - \alpha)e^{-x} - (1 - (1 - \alpha)e^{-x})) + (1 - (1 - \alpha)e^{-x}))$$

$$+ \frac{(1-(1-\alpha)e^{-x})^2}{\alpha e^{-x}} \frac{\alpha}{1-\alpha} \frac{1+(1-\alpha)e^{-x}}{1-(1-\alpha)e^{-x}} (\ln(1 - (1 - \alpha)e^{-x} - \ln\alpha))$$

$$\text{x-g'(x)} = -\frac{1+(1-\alpha)e^{-x}}{1-(1-\alpha)e^{-x}} g(x) \qquad (1.1.17)$$

On simplification, we obtain

$$\frac{x-g\prime(x)}{g(x)} = -\frac{1+(1-\alpha)e^{-x}}{1-(1-\alpha)e^{-x}} \text{ we have}$$

$$\frac{f\prime(x)}{f(x)} = -\frac{1+(1-\alpha)e^{-x}}{1-(1-\alpha)e^{-x}} \qquad (1.1.18)$$

On integrating both sides of (1.1.18), we obtain

$$f(x) = c \frac{\alpha e^{-x}}{(1-(1-\alpha)e^{-x})^2}$$

where c is a constant. Using the boundary condition $F(0) = 0$ and $F(\infty) = 1$, we obtain

$$f(x) = \frac{\alpha e^{-x}}{(1-(1-\alpha)e^{-x})^2}, 0 < x < \infty, \alpha > 0.$$

The following theorem is based on the left truncated first moment.

Theorem 1.1.7.

Suppose the random variable X is continuous (with respect to Lebesgue measure) with cdf F(x), pdf f(x), $F(0) = 0$ and $F(\infty) = 1$. We assume E(x) is finite. Then $E(X|X > x) = h(x)r(x))$ where $r(x) = \frac{f(x)}{1-F(x)}$ and

$$h(x) = \frac{\alpha ln\alpha}{\alpha-1} + x(1-(1-\alpha)e^{-x})$$
$$-\frac{(1-(1-\alpha)e^{-x})^2}{\alpha e^{-x}} \frac{\alpha}{1-\alpha}(ln((1-(1-\alpha)e^{-u}) - ln\alpha)))-, \alpha \neq 1,$$

if and only if

$$f(x) = \frac{\alpha e^{-x}}{(1-ae^{-x})^2}, \alpha > 0, 0 < x < \infty.$$

Proof

Suppose $f(x) = \frac{\alpha e^{-x}}{(1-ae^{-x})^2}, \alpha > 0, 0 < x < \infty.$

Then

$$h(x) = \frac{\int_x^\infty uf(u)du}{f(x)} = \frac{E(x) - \int_0^x uf(u)du}{f(x)}$$

$$= -\frac{\frac{\alpha\ln\alpha}{\alpha-1} - \int_0^x uf(u)du}{f(x)}$$

$$= \frac{\alpha\ln\alpha}{(\alpha-1)f(x)} - - g(x)$$

$$h'(x) = \frac{\alpha\ln\alpha}{(\alpha-1)f(x)} \frac{1+(1-\alpha)e^{-x}}{1-(1-\alpha)e^{-x}} +$$

$$-x - \frac{1+(1-\alpha)e^{-x}}{1-(1-\alpha)e^{-x}} g(x) \qquad \text{(from 1.1.15)}$$

$$= \frac{1+(1-\alpha)e^{-x}}{1-(1-\alpha)e^{-x}} h(x) - x$$

Thus

$$-\frac{x + h'(x)}{h(x)} = -\frac{1+(1-\alpha)e^{-x}}{1-(1-\alpha)e^{-x}}$$

Hence by Lemma A.2 as given in the Appendix, we have

$$\frac{f'(x)}{f(x)} = -\frac{1+(1-\alpha)e^{-x}}{1-(1-\alpha)e^{-x}} \quad (1.1.19)$$

On integrating both sides of (1.1.17), we obtain

$$F(x) = c\frac{\alpha e^{-x}}{(1-(1-\alpha)e^{-x})^2}$$

where c is a constant. Using the boundary condition

$$F(0) = 0$$

and

$F(\infty) = 1,$

we obtain:

$$F(x) = \frac{\alpha e^{-x}}{(1-(1-\alpha)e^{-x})^2}, \, 0 < x < \infty, \alpha > 0.$$

Hossein and Ahsanullah (2013) proved this if the random variable X has the cdf $F(x) = 1 - e^{-x}$, $x \geq 0$, then the relation $E(X^n | X \geq x) = \sum_{j=0}^{k} n^{(j)} x^{n-j} \frac{f(x)}{1-F(x)}$, where $n^{(m)} = n(n-1) \ldots (n-m+1)$.

and $n^{(0)} = 1$ is a characteristic property of the exponential distribution.

The proof is as follows.

It is easy show that the exponential distribution with cdf $F(x)$, $x \geq 0$ satisfies the relation.

Suppose the random variable X is absolutely continuous with cdf $F(x)$ and pdf $f(x)$ defined on $[0,\infty)$. We assume $F(0) = 0$ and $E(X^n)$ exist. The relation $E(X^n | X \geq x) = \sum_{j=0}^{k} n^{(j)} x^{n-j} \frac{f(x)}{1-F(x)}$

Implies that

$$\int_x^\infty u^n f(u) du = f(x) \left(\sum_{j=0}^{k} n^{(j)} x^{n-j} \right)$$

Differentiating both sides of the above equation with respect to x, we obtain

$$-x^n f(x) = f'(x) \sum_{j=0}^{k} n^{(j)} x^{n-j} + f(x) + f(x) \sum_{j=0}^{n-1} n(n-1)^{(i)}).$$

On simplification, we obtain

$$\frac{f'(x)}{f(x).} = -1.$$

On integrating both sides of the above equation and using the boundary conditions $F(0) = 0$ and $F(\infty) = 1$, *we obtain* $F(x) = 1 - e^{-x}, x \geq 0$.

Similarly the relation

$$E(X^n | X \leq x) = (-x^n + e^x \gamma(n, x)) \frac{f(x)}{F(x)}$$

where $\gamma(n, x) = \int_0^x u^{n-1} e^{-u}$ is a characteristic property of the exponential distribution

Proof

It is easy to show that $F(x) = 1 - e^{-x}, x \geq 0$, then $E(X^n | X \leq x) = (-x^n + e^x \gamma(n, x)) \frac{e^x}{1 - e^{-x}}$ Suppose the random variable X is absolutely continuous with cdf $F(x)$ and pdf $f(x)$ defined on $[0, \infty)$. We assume $F(0) = 0$ and $E(X^n)$ exist. The relation $E(X^n | X \geq x) = (-x^n + e^x \gamma(n, x)) \frac{f(x)}{F(x)}$ implies

$$\int_0^x u^n f(u) du = f(x) \ (-x^n + n e^x \gamma(n, x)).$$

Differentiating both sides of the equation, we obtain

$$x^n f(x) = f'(x)(-x^n + n e^x \gamma(n, x)) . + f(x)(-n x^{n-1} + n x^{n-1} + + n e^x \gamma(n, x)).$$

On simplification, we obtain

$$\frac{f'(x)}{f(x).} = -1.$$

On integrating both sides of the above equation and using the boundary conditions $F(0) = 0$ and $F(\infty) = 1$, *we obtain* $F(x) = 1 - e^{-x}, x \geq 0$.

1.2. ORDER STATISTICS

Let $X_1, X_2, \ldots, X_n$ be i.i.d. continuous random variables. Suppose that $F(x)$ be their cumulative distribution function (cdf) and $f(x)$ be the corresponding probability density function (pdf). Let $X_{1,n} \leq X_{2,n} \leq \ldots \leq X_{n,n}$ be the corresponding order statistics. We denote $F_{k,n}(x)$ and $f_{k,n+}(x)$ as the cdf and pdf respectively of $X_{k,n}(x)$, $k = 1, 2, \ldots, n$.

The joint density function of order statistics $X_{1,n}, X_{2,n}, \ldots, X_{n,n}$ has the form

$$f_{1,2,\ldots,n:n}(x_1, x_2, \ldots, x_n) = n! \prod_{k=1}^{n} f(x_k), \quad -\infty < x_1 < x_2 < \ldots < x_n < \infty$$

and

$$f_{1,2,\ldots,n:n}(x_1, x_2, \ldots, x_n) = 0, \text{ otherwise.}$$

The pdf of the $f_{k,n}(x)$ kth order statistics is given as

$$f_{k:n}(x) = \frac{n!}{(k-1)!(n-k)!} (F(x))^{k-1}(1-F(x))^{n-k} f(x),$$

For the standard exponential distribution

$$f_{1,2,\ldots,n:n}(x_1, x_2, \ldots, x_n) = n! e^{-\sum_{j=1}^{n} x_j}, \quad 0 < x_1 < x_2 < \ldots < x_n < \infty.$$

Using the transformation $W_i = (n-i+1)(X_{i,n} - X_{i-1,n})$, $i = 1,2,\ldots,n$ with $X_{0,n}$

$= 0$. We obtain the pdf $f_{1,2,\ldots,n}(w_1,w_2,\ldots,w_n)$ of $W_1, W_2,\ldots,W_n$ as

$$f_{1,2,\ldots,n}(w_1,w_2,\ldots,w_n) = e^{-\sum_{i=1}^{n} w_i}, 0 < w_j < \infty, i = 1,2,\ldots, n$$

Thus W_j, $j = 1,2,\ldots,n$ are i.i.d. exponential with $F(w) = 1 - e^{-w}$, $0 < w < \infty$.

Hence we can write

$$X_{k,n} \stackrel{d}{=} \frac{W_1}{n} + \frac{W_2}{n-1} + \ldots + \frac{W_k}{n-k+1}$$

$$E(X_{k,n}) = E(\frac{W_1}{n} + \frac{W_2}{n-1} + \ldots + \frac{W_k}{n-k+1})$$

$$= \sum_{j=1}^{k} \frac{1}{n-j+1} \; .$$

$$Var(X_{k,n}) = Var(\frac{W_1}{n} + \frac{W_2}{n-1} + \ldots + \frac{W_k}{n-k+1})$$

$$= \sum_{j=1}^{k} \frac{1}{(n-j+1)^2}.$$

$$Cov(X_{k,n}\, X_{j,n}) = \sum_{j=1}^{k} \frac{1}{(n-j+1)^2}. \; 1 \le k < j \le n.$$

There are some simple formulae for distributions of maxima and minima. The pdfs of the smallest and largest order statistics are given respectively as

$$f_{1,n}(x) = n(1 - F(x))^{n-1} f(x)$$

and

$$f_{n,n}(x) = n(F(x))^{n-1} f(x)$$

Mohammad Ahsanullah

The joint pdf $f_{1,n,n}(x,y)$ of $X_{1,n}$ and $X_{n,n}$ is given by

$$f_{1,n,n}(x,y) = n(n-1)\big(F(y)-F(x)\big)^{n-2} f(x)f(y),$$

for $-\infty < x < y < \infty$.

For the standard exponential distribution.

The pdfs $f_{1,n}(x)$ of $X_{1,n}$ and $f_{n.n}(x)$ of $X_{n,n}$ are respectively

$$f_{1,n}(x) = ne^{-nx`}, x \geq 0.$$

and

$$f_{n,n}(x) = n(1 - e^{-x})^{n-1}e^{-x}, x \geq 0.$$

It can be seen that $nX_{1,n}$ has the exponential distribution.,
The pdfs pf $X_{1.n}$ and $X_{n,n}$ are given respectively in Figures 1.2.1 and 1.2.2 for n-3,5 and 10.

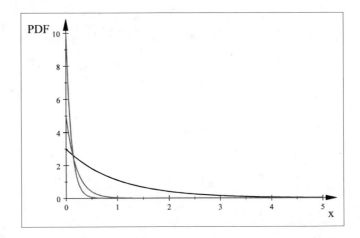

Figure 1.2.1. PDFs $f_{1,3}(x)$ for black, $f_{1,5}(x)$ for red $f_{1.10}(x)$ for -green.

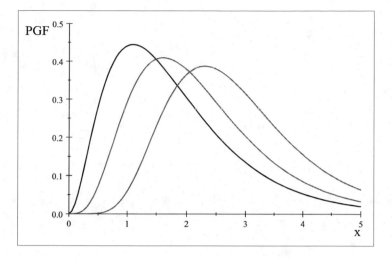

Figure 1.2.2. PDFs $f_{3,3}(x)$ for black, $f_{5,5}(x)$ for red $f_{10,19}(x)$ for green.

The joint pdf $f_{i,j,n}(x,y)$ of $X_{i,n}$ and $X_{j,n}$ is given as

$$f_{i,j,n}(x,y) = \frac{n!}{(i-1)!(j-i-1)!(n-j)!}\left(F(x)\right)^{i-1}\left(F(y)-F(x)\right)^{j-i-1}(1-$$
$$F(y))^{n-j}f(x)f(y)$$
$= 0$, otherwise.

The conditional pdf $f_{j|I,n}(y|x)$ of $X_{j,n}$ given $X_{i,n} = x$ is

$$f_{j|I,n}(y\|x) = \frac{(n-j)!}{(j-i-1)!(n-j)!}(\frac{F(y)-F(x)}{1-F(x)})^{j-i-1}, y>x, \left(\frac{1-F(y)}{1-F(x)}\right)^{n-j}\frac{f(y)}{1-F(x)}$$
$= 0$, otherwise.

Thus $X_{j,n}$ given $X_{i,n} = x$ is the j-ith order statistic in a sample of n-I from truncated distribution with cdf

$$F_c(y|x) = \frac{F(y)-F(x)}{1-F(x).}.$$

For standard exponential distribution with $F(x) = 1-e^{-x}$,

$F_c(y|x) = 1 - e^{-(y-x)}, y > x.$

Let $\alpha_{i,n}^k = E(X_{i,n}^k)$, then we have the following theorems (see Joshi [1978]).

Theorem 1.2.1.

For the standard exponential distribution $\alpha_{i,n}^k = \frac{k}{n}\alpha_{i,n}^{k-1}, k \geq 1, n \geq 1.$

Proof

$$\alpha_{i,n}^k = \int_0^\infty x^k n e^{-nx} dx = -x^k e^{-nx}\big|_0^\infty + \int_0^\infty kx^{k-1}e^{-nx}dx$$
$$= \frac{k}{n}\alpha_{i,n}^{k-1}$$

Theorem 1.2.2.

For the standard exponential distribution,

$$\alpha_{i,n}^{k-1} = \alpha_{i-1,n-1}^k + \frac{k}{n}\alpha_{i,n}^{k-1}, k \geq 1 \text{ and } 2 \leq i \leq n, n \geq 1,$$

Let $\alpha_{i,j,n} = E(X_{i,n}X_{j,n}), 1 \leq i < j \leq n,$

Theorem 1.2.3.

$$\alpha_{i,i+1,n} = \alpha_{i,n}^2 + \frac{1}{n-i}\alpha_{i,n}, 1 \leq i \leq n-1, n \geq 2.$$
$$\alpha_{i,j,n} = \alpha_{i-1,j,n} + \frac{1}{n-j+1}\alpha_{i,j-1,n}, 1 < i < j \leq n, j-i > 2.$$

For proofs of these theorems see Joshi [1978].

It can easily be shown that (for details see Arnold et al. [1973]).

$\sum_{i=1}^{n} \alpha_{i,n}^{1} = nE(X) = n$

$\sum_{i=1}^{n} \alpha_{i,n}^{2} = n((X^2) = 2n$

$E(X_{i,n}) = \sum_{k=1}^{i} \frac{1}{n-k+1}$

$Var(X_{i,n}) = \sum_{k=1}^{i} \frac{1}{(n-k+1)\_2}$

$COV\ (X_{i,n}X_{j,n}) = \sum_{k=1}^{i} \frac{1}{(n-k+1)\_2}, \ 1 \leq i < j \leq n.$

For EEXP with pdf $f_{EEXP} = \alpha e^{-x}(1 - e^{-x})^{\alpha-1}$.

For the MOE with $F(x) = \frac{\alpha e^{-x}}{(1-(1-\alpha)e^{-x})^2}, 0 \leq x < \infty, \alpha > 0$, we have the following theorems for moments of order statistics.

Let $\mu_{r,n}^{k} = E(X_{r,n}^{k}), 1 \leq r \leq n, k > 0$.

Theorem 1.2.4.

$$\mu_{r,n}^{1} = \frac{n!\alpha^{n=r+1}}{(i-1)!(n-i)!} \sum_{j=0}^{\infty} \sum_{i=0}^{r-1} \frac{(-1)^i \binom{r-1}{i}(n+1)}{j!(n-i)!(n-i+1)!} \Phi(n.i,j),$$

where

$$\Phi(n.i,j) = 1 + (n\text{-}r) [\Psi(n-i+1+j) - \Psi(n\text{-}i+j) + \Psi(n-i) - \Psi(n-i+1)]$$

and $\Psi(.)$ Is the digamma function, $\Psi(t) = \frac{d}{dt} ln\Gamma(t)$.

$$\mu_{1,n+1}^{k+1} = \frac{\alpha}{1-\alpha} [\frac{k+1}{n} \mu_{1,n}^{k} - \mu_{1,n}^{k+1}], n \geq 1, k \geq 0, \alpha > 0, \alpha \neq 1.$$

Theorem 1.2.5.

$$\mu_{i+1,n+1}^k = \frac{1}{i}[\frac{(n+1)(k+1)}{n-i+1}\mu_{i,n}^k - \frac{n-\alpha i+1}{\alpha}\mu_{i,n+1}^k$$
$$+\frac{n+1}{\alpha}\mu_{i-1,n]}^{,1<i<n,}, 1\le i\le n, n\ge 1, k>0,$$
$$\mu_{0,r}^k = 0. r > 1, k > 0 \ and \ \mu_{1,r}^0 = 1.$$

For proofs of the theorems 1.2.4 and 1.2.5, see Salah et al [2009].

The following theorem was proved by Bairamov and Ahsanullah (2000).

Theorem 1.2.6.

Let X be an absolutely continuous random variable with cdf F(x) and F(0 = 0. Suppose tha $X_1, X_2,...,X_n$ are independent copies of X and $X_{1,n} < X_{2,n} <...<X_{n,n}$ be the associated order statistics. Then

$$X_{r,n} \overset{d}{=} g_F^{-1}(\sum_{i=1}^r \frac{g_F(x_i)}{n-i+1}, I \le r \le n,$$

where $g_F(x) = -\ln(1 - F(x)$ and g_F^{-1} is the inverse function of $g_F(x)$.

For the exponential distribution with F(x) = 1, $g_F(x) = x$ and $g_F^{-1} = x$.

Thus

$$X_{r,n} \overset{d}{=} \sum_{i=1}^r \frac{x_i}{n-i+1}, 1 \le r \le n.$$

1.3. RECORD VALUES

1.3.1. Definition of Record Values and Record Times

Suppose that $X_1, X_2, \ldots$ is a sequence of independent and identically distributed random variables with distribution function F(x). Let $Y_n = \max(\min)\{X_1, X_2, \ldots, X_n\}$ for $n \geq 1$. We say X_j is an upper (lower) record value of $\{X_n, n \geq 1\}$, if $Y_j > (<)Y_{j-1}, j > 1$. By definition X_1 is an upper as well as a lower record value. One can transform the upper records to lower records by replacing the original sequence of $\{X_j\}$ by $\{-X_j, i \geq 1\}$ or (if $P(X_i > 0) = 1$ for all i) by $\{1/X_i, i \geq 1\}$.

The indices at which the upper record values occur are given by the record times $\{U(n)\}$, $n > 0$, where $U(n) = \min\{j | j > U(n-1), X_j > X_{U(n-1)}, n > 1\}$ and $U(1) = 1$ The record times of the sequence $\{X_n \; n \geq 1\}$ are the same as those for the sequence $\{F(X_n), n \geq 1\}$. Since F(X) has an uniform distribution, it follows that the distribution of U(n), $n \geq 1$, does not depend on F. We will denote L(n) as the indices where the lower record values occur. By our assumption $U(1) = L(1) = 1$. The distribution of L(n) also does not depend on F.

1.3.2. The Exact Distribution of Record Values
for Continuous Random Variables

Many properties of the record value sequence can be expressed in terms of the function R(x), where $R(x) = -\ln \overline{F}(x)$, $0 < \overline{F}(x) < 1$ and $\overline{F}(x) = 1 - F(x)$. Here 'ln' is used for the natural logarithm. We will denote the n-th upper record values as X(n)

If we define $F_n(x)$ as the distribution function of the nth upper record $X(n)$ for $n \geq 1$, then we have for details, see Ahsanullah [1988b].

$F_1(x) = P [X (1) \leq x] = F(x),$

$F_2(x) = P[X (2) \leq x]$

$= \int_{-\infty}^{x} \int_{-\infty}^{y} \frac{dF(u)}{1-F(u)} dF(y)$

$= \int_{-\infty}^{x} R(y)dF(y)$

If $F(x)$ has a density $f(x)$, then the pdf $f_2(x)$ is

$f_2(x) = R(x)f(x)$

The cdf $F_n(x)$ of $X(n)$ is

$F_n(x) = P(X(n) \leq \_x)$

$= \int_{-\infty}^{x} f(u_n)du_n \int_{-\infty}^{u_n} \frac{f(u_{n-1})}{1-F(u_{n-1})} du_{n-1} \dots \int_{-\infty}^{u_2} \frac{f(u_1)}{1-F(u_1)} du_1$

$= \int_{-\infty}^{x} \frac{R^{n-1}(u)}{(n-1)!} dF(u), \quad -\infty < x < \infty .$ \hfill (1.3.2.1)

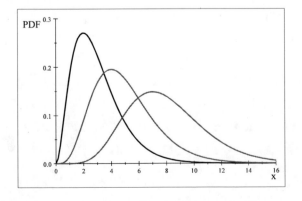

Figure 1.3.1. PDFs- $f_3(x)$ for black, $f_5(x)$ for red and $f_8(x)$ for green.

This can be expressed as

$$F_n(x) = \int_{-\infty}^{R(x)} \frac{u^{n-1}}{(n-1)!} e^{-u} \, du, \quad -\infty < x < \infty \ .$$

$$\bar{F}_n(x) = 1 - F_n(x) = \bar{F}(x) \sum_{j=0}^{n-1} \frac{(R(x))^j}{j!} = e^{-R(x)} \sum_{j=0}^{n-1} \frac{(R(x))^j}{j!}.$$

The pdf $f_n(x)$ of $X(n)$ is

$$f_n(x) = \frac{R^{n-1}(x)}{(n-1)!} f(x), \qquad -\infty < x < \infty \ .$$

The pdf of $X(n)$ if the observations are from an exponential distribution with $F(x) = 1 - \exp(-x), x > 0$ is as follows.

$$f_n(x) = \frac{x^{n-1}}{\Gamma(n)} e^{-x}, x \geq 0 \ .$$

Figure 1.3.1 gives the pdf' of $X(n)$ for $n = 3, 5$ and 8.

The joint pdf of $X(m)$ and $X(n)$ is

$$f_{m,n}(x, y) = \frac{x^{n-1}}{\Gamma(n)\Gamma(n-m)} (y - x)^{n-m-1} e^{-y}, 0 < x < y < \infty.$$

1.3.3. Conditional Distribution

The conditional pdf of $X(j)$ under the condition $X(i) = x_i$ is

$$f(x_j \mid X(i) = x_i) = \frac{f_{ij}(x_i, x_j)}{f_i(x_i)}$$

$$= \frac{(R(x_j) - R(x_i))^{j-i-1}}{(j-i-1)!} \frac{f(x_j)}{1 - F(x_i)} \tag{1.3.3.1}$$

for $-\infty < x_i < x_j < \infty$.

For $i > 0$, $1 \le k < m$, the joint conditional pdf of $X(i + k)$ and $X(i + m)$ under the condition $X(i) = z$ is

$$f_{i+k, i+m}(x, y \mid X(i) = z)$$
$$= \frac{1}{\Gamma(m-k)} \cdot \frac{1}{\Gamma(k)} \cdot [R(y) - R(x)]^{m-k-1} [R(X)$$
$$- R(z)]^{k-1} \frac{f(y) r(x)}{\bar{F}(z)}$$

for $-\infty < z < x < y < \infty$.

If the observations are from an exponential distribution with cdf $F(x)$ = 1-exp(-x), $x \ge 0$, then we have $R(x) = x$ and the joint pdf of $X(m)$ and $X(n)$, $n > m$, is

$$f_{m,n}(x,y) = \frac{x^{m-1}}{\Gamma(m)\,\Gamma(n-m)} (y-x)^{n-m-1} \, e^{-y} \,, 0 \le x < y < \infty,$$

$$= 0, \text{ otherwise.}$$

The conditional pdf of $X(n)$ under the condition $X(m) = x$ is

$$f(y \mid X(m) = x) = \frac{(y-x)^{n-m-1}}{\Gamma(n-m)} e^{-(y-x)} \,, 0 \le x < y < \infty$$

$$= 0, \text{ otherwise.}$$

It can easily be seen that that $X(n) - X(m)$ and $X(m)$) are independent for $n > m$ if $F(x) = 1-e^{-x}$, $x \geq 0$.

For exponential distribution with $F(x) = 1-\exp(-x)$, $x \geq 0$,

$E(X(n)) = n$

$Var\ (X(n)) = n$

and COV $(X(m), X(n)) = m$, if $m < n$.

The joint pdf $f(x_1, x_2, \ldots, x_n)$ of the n record values $X(1), X(2), \ldots, X(n)$ is given by

$f(x_1, x_2, \ldots, x_n) = r(x_1)r(x_2) \ldots f(x_n)$,

- for $-\infty < x_1 < x_2 < \ldots < x_{n-1} < x_n < \infty$,

for $-\infty < x_1 < x_2 < \ldots < x_n < \infty$

where $r(x) = \dfrac{d}{dx} R(x) = \dfrac{f(x)}{1 - F(x)}$, $0 < F(x) < 1$.

The function $r(x)$ is known as hazard rate.

The joint pdf of $X(i)$ and $X(j)$ is

$$f(x_i, x_j) = \frac{(R(x_i))^{i-1}}{(i-1)!} r(x_i) \frac{(R(x_j) - R(x_i))^{j-i-1}}{(j-i-1)!} f(x_j)$$

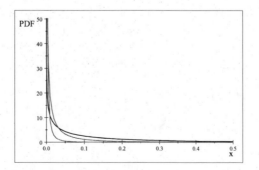

Figure 1.3.2. PDFs- $f_{(3)}(x)$ for back, $f_{(5)}(x)$ for red and $f_{(8)}(x)$ for -green.

The marginal pdf of the n-th lower record value can be derived by using the same procedure as that of the n-th upper record value. We use $H(u) = -\ln F(u)$, $0 < F(u) < 1$ and $h(u) = -\dfrac{d}{du} H(u)$, We will denote $X_{(m)}$ as the m-th lower record value.

$$P(X_{(n)} \leq x) = \int_{-\infty}^{x} \frac{\{H(u)\}^{n-1}}{(n-1)!} dF(u)$$

and the corresponding pdf $f_{(n)}$ can be written as

$$f_{(n)}(x) = \frac{(H(x))^{n-1}}{(n-1)!} f(x).$$

For exponential distribution with $F(x) = 1-e^{-x}$, we have

$$f_{(n)}(x) = \frac{(-\ln(1-e^{-x})^{n-1}}{\Gamma(n)} e^{-x}, x > 0.$$

Figure 1.3.2 gives the pdf of $f_{(n)}(x)$ on $= 3, 5$ and 9.
The joint pdf of $X_{(1)}, X_{(2)}, \ldots, X_{(n)}$ can be written as

$$f_{(1),(2),\ldots,(n)}(x_1, x_2, \ldots, x_n) = h(x_1)h(x_2)\ldots h(x_{n-1})f(x_n), \quad -\infty < x_n < x_{n-1}$$
$$< \ldots < x_1 < \infty,$$
$$= 0, \text{ otherwise.}$$

The joint pdf of $X_{(r)}$ and $X_{(s)}$ is

$$f_{(r),(s)}(x,y) = \frac{(H(x))^{r-1}}{(r-1)!} \frac{[H(y)-H(x)]^{s-r-1}}{(s-r-1)!} h(x)f(y)$$

for s > r and $-\infty < y < x < \infty$.

Proceeding as in the case of upper record values, we can obtain the conditional pdfs of the lower record values. The conditional pdf of the jth lower record $X_{(j)}$ under the condition that the ith lower record $X_{(i)} = x_i$ is given is

$$f(\,x_j\,|\,X_{(i)} = x_i\,) = \frac{f_{ij}(x_i,x_j)}{f_i(x_i)}$$

$$= \frac{(H(x_j)-H(x_i))^{j-i-1}}{(j-i-1)!}\frac{f(x_j)}{F(x_i)},$$

for $-\infty < x_j < x_i < \infty$.

Theorem 1.3.1.

For the exponential distribution with $F(x) = 1 - e^{-x}, x > 0,$ Let $\mu_n^p = E(X(n))^p,$ then $\mu_n^p = \frac{\Gamma(n`+p)}{\Gamma(n)}.$

Proof

$$\mu_n^p = \frac{\int_0^\infty x^k e^{0x}dx}{\Gamma(n)} = \frac{\Gamma(n`+p)}{\Gamma(n)}.$$

Theorem 1.3.2.

For $n \geq 1$ and r = 0,1,2,...

$$E((X(n))^{r+1} = E((X(n-1))^{r+1} + (r+1)\,E((X(n))^r$$

with $E((X(0))^{r+1} = 0$ and $E((X(n))^0 = 1.$

For proof see Balakrishnan and Ahsanullah (1995).

1.3.4. Representations of Records

1.3.4.1. Representations of Upper Records

Let Y_1, Y_2,...., Y_n,... be a sequence of independent and identically distributed random variables with the cdf $F_0(x) = 1 - \exp(-x)$, $x > 0$. Further suppose that X_1, X_2,... be a sequence of i.i.d. r.v.'s with some continuous cdf F. Then one has that

$$- \ln(1 - F(X_k)) \overset{d}{=} Y_k, \, k = 1, 2, \ldots.$$

The following theorem (Bairamov and Ahsanullah [2000] gives the representation of the n-th record as the sum of n independent random variables.

Theorem 1.3.1. The Following Equality Holds

$$X(n) \overset{d}{=} g_F^{-1}(g_F(X_1) + g_F(X_2) + \ldots + g_F(X_n)) \, ,$$

where $g_F(x) = -\ln(1 - F(x))$ and $g_F^{-1}(x) = F^{-1}(1 - e^{-x})$.

Proof

$$P(X(n) \le x) = P(Y_1 + Y_2 + \ldots + Y_n \le -\ln(1 - F(x)))$$
$$= P(-\ln(1 - F(X_1)) + (-\ln(1 - F(X_2)) + \ldots +$$
$$(-\ln(1 - F(x_n)) \le -\ln(1 - F(x)))$$
$$= P(g_F(X_1) + g_F(X_2) + \ldots + g_F(X_n) \le g_F(x)),$$

where $g_F(x) = -\ln(1 - F(x))$, and hence

$$P(X(n) \leq x) = P(g_F^{-1} \{g_F(X_1) + g_F(X_2) + \ldots + g_F(X_n)\} \leq x),$$

where $g_F^{-1}(x) = F^{-1}(1-e^{-x})$.

It means that

$$X(n) \overset{d}{=} g_F^{-1}(g_F(X_1) + g_F(X_2) + \ldots + g_F(X_n)).$$

For the exponential distribution with cdf F(x) as $F(x) = 1-e^{-x}$ $x \geq 0$, it is possible to show that

$$X(n) \overset{d}{=} X_1 + X_2 + \ldots + X_n .$$

Proof

We have:

$$g(x) = -\ln(1-F(x)) = x ,$$

$$\sum_{k=0}^{n} g(X_i) = \sum_{k=0}^{n} X_i ,$$

$$F^{-1}(x) = -\ln(1-x) \text{ and } F^{-1}(1 - e^{-x}) = x ,$$

and

$$F^{-1}\left(1 - e^{-\sum_{k=1}^{n} g(X_i)}\right) = F^{-1}(1-e^{-\sum_{k=1}^{n} X_i}) = \sum_{k=1}^{n} X_i.$$

Thus,

$$X(n) \overset{d}{=} X_1 + X_2 + \ldots + X_n ,$$

where X_i, i = 1, 2, ..,n, are i.i.d r.v.'s with d.f. $F(x) = 1-e^{-x}$, $x \geq 0$.

1.3.4.2. Representation of Lower Records

Suppose the random variable X has the cdf F(x) and $F^{-1}(x)$ is the inverse of the cdf F(x). Kirmani and Ahsanullah [2019] proved that the f_n (x) the pdf of the nth lower record is distributed as -ln $(1-e^{-X(n)})$, where X(n) is the nth upper record of X. Let L_n be the nth lower record from the exponential distribution with $F(x) = 1-e^{-x}$, $x \geq 0$.then

$$E(L_n) = E(-\ln (1-e^{-X(n)}))$$

$$= E(- \sum_{k=1}^{\infty} (-1)^{k+1} \frac{(e^{-X(n)})^k}{k})$$

$$= \sum_{k=1}^{\infty} \frac{E(e^{-X(n)})^k}{k}$$

$$= \sum_{k=1}^{\infty} \frac{1}{k}(\frac{1}{k+1})^n$$

$$= E(L_{n-1}) - \sum_{k=2}^{\infty} (\frac{1}{k})^n, \text{ n} \geq 2.$$

$$= E(L_{n-1}) - \varsigma(n) + 1$$

where $\varsigma(n) = \sum_{k=1}^{\infty} (\frac{1}{k})^n$

1.4. GENERALIZED ORDER STATISTICS

Kamps [1995] introduced the concept of the generalized order statistics *(gos)* as general framework models of ordered random variables. Sequential order statistics, upper record valuess, progressively Type-II censored order statistics and some other ordered random variables can be considered as a special case of the *gos*. These models can be effectively applied, e.g., in reliability theory. Although the *gos* contains many useful models of ordered random variables, the random variables that are decreasingly ordered cannot be integrated into this frame. Consequently, this model is inappropriate to study for decreasing order statistics. Using the concept gos, Burkschat et al. [2003] introduced the concept of dual generalized order statistics (dgos) as a systematic approach to some models of decreasingly ordered random variables. Dual generalized order statistics represents a unification of models of decreasingly ordered random variables e.g., reversed order statistics, lower records, lower k- records and lower Pfeifer records.

Definitions

Suppose $X(1,n,m,k)$, ..., $X(n,n,m,k)$, ($k \geq 1$, m is a real number), are n generalized order statistics. Then the joint pdf $f_{1,...,n}(x_1,...,x_n)$ can be written as

$$f_{1,...,n}(x_1,...,x_n) = k$$
$$\prod_{j=1}^{n-1} \gamma_j \prod_{i=1}^{n-1} (1-F(x_i))^m f(x_i)(1-F(x_n))^{k-1} f(x_n), \quad (1.4.1)$$

for $F^{-1}(0) < x_1 < \cdots < x_n < F^{-1}(1)$,
= 0, otherwise,

where $\gamma_j = k + (n-j)(m+1)$ and $f(x) = \dfrac{dF(x)}{dx}$.

If $m = 0$ and $k = 1$, then $X(r,n,m,k)$ reduces to the ordinary rth order statistic and (1.4..1) is the joint pdf of the n order statistics $X_{1,n} \leq ... \leq X_{n,n}$. If $k = 1$ and $m = -1$, then (1.4.1) is the joint pdf of the first n upper record values of the independent and identically distributed random variables with distribution function $F(x)$ and the corresponding probability density function $f(x)$.

Let $F_{r,n,m,k}(x)$ be the distribution function of $X(r,n,m,k)$. Then the following equalities are equal:

$$F_{r,n,m,k}(x) = I_{\alpha(x)}(r, \frac{\gamma r}{m+1}), \text{ if } m > -1$$

and

$$F_{r,n,m,k}(x) = \Gamma_{\beta(x)}(r), \text{ if } m = -1,$$

where

$$\alpha(x) = 1 - (\overline{F}(x))^{m+1}, \overline{F}(x) = 1 - F(x)$$
$$\beta(x) = -k \ln \overline{F}(x)),$$
$$I_x(p,q) = \frac{1}{B(p,q)} \int_0^x u^{p-1}(1-u)^{q-1} du, B(p,q) = \frac{\Gamma(p)\Gamma(q)}{\Gamma(p+q)},$$

and

$$\Gamma_x(r) = \int_0^x \frac{1}{\Gamma(r)} u^{r-1} e^{-u} du.$$

For m > -1, it follows from (20.1) that

$$F_{r,n,m,k}(x) = \int_{F^{-1}(0)}^x \frac{c_r}{(r-1)!} (1-F(u))^{k+(n-r)(m+1)-1} g_m^{r-1}(F(u)) f(u) \, du$$

Using the relation

$$B(r, \frac{\gamma r}{m+1}) = \frac{\Gamma(r)(m+1)^r}{c_r}$$

and substituting $t = 1 - (\bar{F}(x))^{m+1}$, we get on simplification

$$F_{r,n,m,k}(x)$$

$$= \frac{1}{B(r, \frac{\gamma_r}{m+1})} \int_0^1 (\bar{F}(x))^{m+1} \; (1-u))^{\gamma_r - 1}(1-u)^{r-1} du$$

$$= I_{\alpha(x)}\left(r, \frac{\gamma_r}{m+1}\right). \tag{1.4.2}$$

For m = -1 the following relations are valid:

$$F_{r,n,m,k}(x) = \int_{F^{-1}(0)}^{x} \frac{k^r}{(r-1)!}(1-F(u))^{k-1}(-\ln(1-F(u))^{r-1}f(u)du$$

$$= \int_0^{-k \ln \bar{F}(x)} \frac{1}{(r-1)!} t^{r-1} e^{-t} \, dt$$

$$= \Gamma_{\beta(x)}(r), \quad \beta(x) = -k \ln \bar{F}(x).$$

The following equality is valid:

$$F_{r,n,m,k}\left(r, \frac{\lambda_r}{m+1}\right) - F_{r,n,m,k}\left(r+1, \frac{\lambda_{r+1}}{m+1}\right) = \frac{1}{\gamma_{r+1}} \frac{\bar{F}(x)}{f(x)} f_{r+1,n,m,k} \, .$$

Let X be a random variable (r.v.) whose probability density function (pdf) f is given by

$$f(x) = \sigma^{-1} \exp(-\sigma^{-1}x), \text{ for } x > 0, \sigma > 0,$$
$$= 0, \text{otherwise}.$$

then

$$X(1, n, m, k) \stackrel{d}{=} E(0, \frac{\sigma}{\gamma_1})$$

Integrating out $x_1, \ldots, x_{r-1}, x_{r+1}, \ldots, x_n$ from (1.3.1), we get the pdf $f_{r,n,m,k}$ of $X(r,n,m,k)$ as

$$f_{r,n,m,k}(x) = \frac{c_r}{(r-1)!}(1 - F(u))^{k+(n-r)(m+1)-1} g_m^{r-1}(F(u))f(u)'$$

where

$$c_r = \prod_{j=1}^{r} \gamma_j$$

and

$$g_m(x) = \frac{1}{m+1}(1 - (1-x)^{m+1}), m \neq -1,$$
$$= -\ln(1-x), m = -1, x \in (0,1).$$

Since

$$\lim_{m \to -1} \frac{1}{m+1}(1 - (1-x)^{m+1}) = -\ln(1-x),$$

we will write

$$g_m(x) = \frac{1}{m+1}(1 - (1-x)^{m+1}), \quad \text{for all } x \in (0,1) \text{ and for all m with}$$
$$g_{-1}(x) = \lim_{m \to -1} g_m(x).$$

Substituting $F(x) = 1 - e^{-x}$, we obtain for the standard exponential distribution when $r = 1$,

$$f_{1,n,m,k}(x) = \frac{\gamma_1}{\sigma} e^{-\gamma_1 x / \sigma}, \; x > 0, \sigma > 0.$$

Thus,

$$X(1, n, m, k) \overset{d}{=} E(0, \frac{\sigma}{\gamma_1}).$$

The moment generating function $M_{r,n,m,k}(t)$ of $X(r,n,m,k)$ is

$$M_{r,n,m,k}(t) = \int_0^\infty e^{tx} \frac{c_r}{(r-1)!} e^{-\gamma_r \sigma^{-1} x} \left\{ \frac{1}{m+1} \left[1 - e^{-(m-1)\gamma_r \sigma^{-1} x)} \right] \right\}^{r-1} dx$$

$$= \frac{c_r}{(r-1)!} \int_0^\infty e^{-y(\gamma_r - t\sigma)} \left\{ \frac{1}{m+1} \left[1 - e^{-(m+1)y} \right] \right\}^{r-1} dy.$$

Using the following property (see Gradsheyn and Ryzhik [1965], p. 305)

$$\int_0^\infty e^{-ay} (1 - e^{-by})^{r-1} dy = \frac{1}{b} B(\frac{a}{b}, r) \text{ and } \gamma_r + i(m+1) = \gamma_{r-i},$$

From above we have

$$M_{r,n,m,k}(t) = \frac{c_r}{(r-1)!} \frac{(r-1)!}{\prod_{j=1}^r \gamma_j \left(1 - \frac{t\sigma}{\gamma_j} \right)} = \prod_{j=1}^r \left\{ 1 - \frac{t\sigma}{\gamma_j} \right\}^{-1}.$$

Thus,

$$X(r + 1, n, m, k) \stackrel{d}{=} X(r, n, m, k) + \sigma \frac{W}{\gamma_{r+1}}. \tag{1.4.3}$$

It follows from (1.4.3) that

$$X(r + 1, n, m, k) \stackrel{d}{=} \sigma \sum_{j=1}^{r+1} \frac{W_j}{\gamma_j},$$

where $W_1, ..., W_{r+1}$ are i.i.d. with $W_i \stackrel{d}{=} E(0,1)$. If we take $k = 1$, $m = 0$, then we obtain from (1.4.2) the well-known result (see Ahsanullah and Nevzorov [2005, p.51] for order statistics for $X \in E(\sigma)$:

$$X_{r+1,n} \stackrel{d}{=} \sigma \sum_{j=1}^{r+1} W_j/(n-j+1)$$

From (1.4.3) it follows that $\gamma_{r+1}\{X(r + 1,n,m,k) - X(r,n,m,k)\}$ is distributed as exponential. This property can also be obtained by considering the joint pdf of $X(r + 1,n,m,k)$ and $X(r,n,m,k)$, using the transformations $U_1 = X(r,n,m,k)$ and $D_{r+1} = \gamma_{r+1}(X(r + 1,n,m,k) - X(r,n,m,k))$ and integrating with respect to U_1.

For the exponential distribution with $F(x) = 1 - e^{-x}, x \geq 0$, $X(r,n,m,k)$- $X(r-1,n,m,k)$ and $X(r-1,n,m,k)$, $1 < r \leq n$, are independent.

Kamps and Cramer [2001] proved that for the exponential distribution with $F(x) = 1 - e^{-x}, x \geq 0$,

$$E((X(r, n, m, k)^l - (E((X(r-1, n, m, k)^l = \frac{l}{\lambda_r} E((X(r, n, m, k)^{l-1}$$

Beg et. al. (2013) has shown that rth gos can be represented as

$$X(r, n, m, k) \overset{d}{=} g_F^{-1}\left(\sum_{i=1}^{r} \frac{g_F(X_i)}{\gamma_i}\right), 1 \leq r \leq n,$$

where $g_F(x) = -\ln(1 - F(x))$, $g_F^{-1}(x) = F^{-1}(1-e^{-1})$.

If $F(x) = 1-e^{-x}$, then $X * r, n, m, k) \overset{d}{=} \sum_{i=1}^{n} \frac{W_i}{\gamma_i}$,

where $W_1, W_2, \ldots, W_n$ are i.i.d with cdf $F(x) = 1-e^{-x}$.

1.4.1. Dual Generalized Order Statistics (dgos)

Suppose $X^*(1, n, m, k), X^*(2, n, m, k), \ldots, X^*(n, n, m, k)$ are n lower generalized order statistics from a continuous cumulative distribution function (cdf) $F(x)$ with the corresponding probability density function (pdf) f(x). Their joint pdf $f *_{12\ldots n}(x_1, x_2, \ldots, x_n)$ is given by

$$f^*_{12\ldots n}(x_1, x_2, \ldots, x_n) = k \prod_{j=1}^{n-1} \gamma_j \prod_{i=1}^{n-1} (F(x_i))^m (F(x_n))^{k-1} f^{(x)}$$

for $\overset{-1}{F}(1) \geq x_1 \geq x_2 \geq \ldots \geq \overset{-1}{F}(0), m \geq -1$, where $\gamma_r = k + (n - r)$ $(m + 1), r = 1, 2, \ldots, n - 1, k \geq 1$ and n is a positive integer.

The marginal pdf of the rth lower generalized order (dgos) statistics is

$$f^*_{r,n,m,k}(x) = \frac{c_{r-1}}{\Gamma(r)} (F(x))^{\gamma_r - 1} (g_m F(x))^{r-1} f(x),$$

where

$$c_{r-1} = \prod_{i=1}^{r} \gamma_i,$$

$$g_m(x) = \frac{1}{m+1}\left(1 - x^{m+1}\right), \text{ for } m \neq -1,$$

and

$$g_m(x) = -\ln x, \text{ for } m = -1.$$

Since $\lim_{m \to -1} g_m(x) = -\ln x,$

we will take $g_m(x) = \frac{1}{m+1}\left(1 - x^{m+1}\right)$ for all m with $g_{-1}(x) = -\ln x$. For $m = 0, k = 1$, $X^*(r,n,m,k)$ reduces to the order statistics $X_{n-r+1,n}$ from the sample $X_1, X_2, \dots, X_n$. If $m = -1$, then $X^*(r,n,m,k)$ reduces to the rth lower k- record value.

If F(x) is continuous, then

$$\bar{F}^*_{r,n,m,k}(x) = 1 - F^*_{r,n,m,k}(x) = I_{\alpha(x)}\left(r, \frac{\gamma_r}{m+1}\right),$$

if $m > -1$,

$$= \Gamma_{\beta(x)}(r), \text{ if } m = -1,$$

where

$$\alpha(x) = 1 - \left(F(x)\right)^{m+1}, \; I_x(p,q) = \frac{1}{B(p,q)} \int_0^x u^{p-1}(1-u)^{q-1} du,$$

$x \leq 1,$

$$\beta(x) = -k\ln F(x), \; \Gamma_x(r) = \frac{1}{\Gamma(r)} \int_0^x u^{r-1}e^{-u} du$$

and

$$B(p,q) = \frac{\Gamma(p)\Gamma(q)}{\Gamma(p+q)}.$$

For $m > -1$ we get that

$$1-F^*_{r,n,m,k}(x) = \frac{c_{r-1}}{\Gamma(r)} \int_x^\infty (F(u))^{\gamma_r-1}(g_m(F(u)))^{r-1} f(u)du$$

$$= \frac{c_{r-1}}{\Gamma(r)} \int_x^\infty (F(u))^{\gamma_r-1}\left[\frac{1-(F(u))^{m+1}}{m+1}\right]^{r-1} f(u)du$$

$$= \frac{c_{r-1}}{\Gamma(r)}\frac{1}{(m+1)^r}\int_0^{1-(F(x))^{m+1}} t^{r-1}(1-t)^{(\gamma_r/(m+1))-1}dt$$

$$= I_{\alpha(x)}\left(r,\frac{\gamma_r}{m+1}\right).$$

For $m = -1, \gamma_j = k, j = 1,2,\ldots,n$ and

$$1-F^*_{r,n,m,k}(x) = \int_x^\infty \frac{k^r}{\Gamma(r)}(F(u))^{k-1}(-\ln F(u))^{r-1} f(u)du$$

$$= \int_0^{-k\ln F(x)} \frac{t^{r-1}e^{-t}}{\Gamma(r)}dt$$

$$= \Gamma_{\beta(x)}(r), \beta(x) = -k\ln F(x).$$

Suppose that X is an absolutely continuous random variable with cdf F(X) and pdf f(x). The following relations are true.

$$\gamma_{r+1}\left(F^*_{r+1,n,m,k}(x) - F^*_{r,n,m,k}(x)\right) = \frac{F(x)}{f(x)}f^*_{r+1,n,m,k}(x),$$

for $m > -1$,

and

$$k\left(F^*_{r+1,n,m,k}(x) - F^*_{r,n,m,k}(x)\right) = \frac{F(x)}{f(x)} f^*_{r+1,n,m,k}(x),$$

for $m = -1$.

For $m > -1$ the following equalities are valid:

$$F^*_{r+1,n,m,k}(x) - F^*_{r,n,m,k}(x)$$
$$= I_{\alpha(x)}\left(r, \frac{\gamma_r}{m+1}\right) - I_{\alpha(x)}\left(r+1, \frac{\gamma_{r+1}}{m+1}\right)$$
$$= I_{\alpha(x)}\left(r, \frac{\gamma_r}{m+1}\right) - I_{\alpha(x)}\left(r+1, \frac{\gamma_r}{m+1} - 1\right).$$

We know that

$$I_x(a, b) - I_x(a+1, b-1) = \frac{\Gamma(a+b)}{\Gamma(a+1)\Gamma(b)} x^a (1-b)^{b-1}.$$

Thus,

$$F^*_{r+1,n,m,k}(x) - F^*_{r,n,m,k}(x)$$
$$= \frac{\Gamma(r + \frac{\gamma_r}{m+1})}{\Gamma(r+1)\Gamma(\frac{\gamma_r}{m+1})} \left(1 - (F(x))^{m+1}\right)^r \left(F(x)^{m+1}\right)^{\frac{\gamma_r}{m+1} - 1}$$
$$= \frac{\gamma_1 \cdots \gamma_r}{\Gamma(r+1)} \left(\frac{1 - (F(x))^{m+1}}{m+1}\right)^r (F(x))^{\gamma_{r+1}}$$
$$= \frac{F(x)}{\gamma_{r+1} f(x)} f^*_{r+1,n,m,k}(x).$$

Hence,

$$\gamma_{r+1}\left(F^*_{r+1,n,m,k}(x) - F^*_{r,n,m,k}(x)\right) = f^*_{r+1,n,m,k}(x) \frac{F(x)}{f(x)}.$$

For $m = -1$ one gets that

$$F^*_{r+1,n,m,k}(x) - F^*_{r,n,m,k}(x) = \Gamma_{\beta(x)}(r) - \Gamma_{\beta(x)}(r+1),$$

where $\beta(x) = $ -kln F(x).

Hence

$$F^*_{r+1,n,m,k}(x) - F^*_{r,n,m,k}(x)$$

$$= (\beta(x))^r e^{-\beta(x)} \frac{1}{\Gamma(r+1)}$$

$$= \frac{(F(x))^k}{\Gamma(r+1)} (-k \ln F(x))^r.$$

Thus,

$$k\left[F^*_{r+1,n,m,k}(x) - F^*_{r,n,m,k}(x)\right] = \frac{F(x)}{f(x)} f^*_{r+1,n,m,k}(x).$$

$$F_Y(x) = P(Y \leq x) = P(X^*(r,n,m,k)W_{r+1} \leq x).$$

Chapter 2

CHARACTERIZATIONS
BY ORDER STATISTICS

In this chapter we will present some characterizations of the exponential distribution by various properties of order statistics.

2.1. CHARACTERIZATIONS
BY CONDITIONAL EXPECTATIONS

We assume that the random variable X is continuous with cdf $F(x)$ and pdf $f(x)$. We assume further that $E(X)$ exists. Consider the relation

$$E(X_{j,n}|X_{i,n} = x) = ax + b, \ 1 \le i < j \le n. \qquad (2.1.1).$$

Fisz [1958] considered the characterization of the exponential distribution by considering $j = 2$, $j = 1$ and $a = 1$. Roger (1963) characterized the exponential distribution by considering $j = i + 1$ and $a = 1$. Ferguson [1963] gave the characterization of exponential, Pareto and power function distribution by considering j = i + 1 and with different values of a = 1, a > 1 and a < 1.

The following theorem was given by Gupta and Ahsanullah [2004].

Theorem 2.1.1.

Under some mild conditions on $\varphi(x)$ and g(x) the relation

$$E(\varphi(X_{i+s,n})|X_{i,n} = x) = g(x)\, 0 < s \leq n - i \qquad (2.1.2)$$

determines the distribution F(x) uniquely.

If s = 1, $\varphi(x) = x$ and g(x) = ax + b, then the relation (2.1.1) will lead to the equation

$$r(x) = \frac{g\prime(x)}{(n-i)(a-1)x + b,} \qquad (2.1.3)$$

where r(x) is the hazard rate and $r(x) = \frac{f(x)}{1-F(x)}$.

If a = 1, then we obtain

$$F(x) = 1 - e^{-\lambda(x-\mu)} , \qquad (2.1.4)$$

for x ≥ μ and $\lambda = \frac{a}{b(n-1)}$.

Wesolowski and Ahsanullah [2001] extended the result of Ferguson. They gave the following theorem.

Theorem 2.1.2.

Suppose that X is an absolutely continuous random variable with cdf F(x) and pdf f(x). If $E(X_{i+2,n})$ is finite for $1 < i < n-2$, $n > 2$, then $E(X_{i+2} | X_{i,n} = x) = x + b$ iff

$$F(x) = 1 - e^{-\lambda(x-\mu)} , x \geq \mu \text{ and } \lambda = \frac{2n-2i-1}{b(n-i)(n-i-1)}.$$

Deminska and Wesolowski [1998] gave the following general result.

Theorem 2.1.3.

Suppose that X is an absolutely continuous random variable with cdf F(x) and pdf f(x). If $E(X_{i+s,n})$ is finite for $1 \leq 1 \leq i < n-s$, $s \geq 1$, $n > 2$, then

$$E(X_{i+s} | X_{i,n} = x) = x + b \text{ iff}$$

$$F(x) = 1 - e^{-\lambda(x-\mu)} , x > \mu \text{ and } b = \frac{(n-i)!}{\lambda(n-s-i)!} \sum_{m=0}^{s-1} \frac{(-1)^m}{m!(n-i-s+1+m)^2}.$$

The following theorem gives a characterization of the exponential distribution based on the conditional expectation of the smallest order statistics.

Theorem 2.1.4.

Suppose that X_i, $i = ,2,\ldots,n$ are independent and identically distributed random variables with cdf F(x) and pdf f(x). We assume that E(x) exists. Then $E(\bar{X} | X_{1,n} = x) = x + b$, where $b = \frac{n-1}{n}$ and $\bar{X} = \frac{1}{n}(X_1 + X_2 + .. + X_n)$ iff

$F(x) = 1 - e^{-x}$, $x > 0$.

Proof

$E(X_i | X_{1,n} = x) = \frac{x}{n} + \frac{n-1}{n} \frac{\int_x^\infty y f(y) dy}{1 - F(x)}$.

If $F(x) = 1 - exp(-x)$, then

$E(X_i | X_{1,n} = x) = \frac{x}{n} + \frac{n-1}{n} \frac{\int_x^\infty y e^{-y} dy}{e^{-x}}$.

$= = \frac{x}{n} + \frac{n-1}{n}(x + 1)$

$= x + b, \ b = \frac{n-1}{n}$.

Thus

$E(\bar{X} | X_{1,n} = x) = E(X_i | X_{1,n} = x) = x + b.$

Suppose that

$E(\bar{X} | X_{1,n} = x) = E(X_i | X_{1,n} = x) = x + b.$

Since $E(\bar{X} | X_{1,n} = x) = E(X_i | X_{1,n} = x)$, we have

$= \frac{x}{n} + \frac{n-1}{n} \frac{\int_x^\infty y f(y) dy}{1 - F(x)} = x + \frac{n-1}{n}$

i.e.:

$\frac{\int_x^\infty y f(y) dy}{1 - F(x)} = x + 1.$

Thus

$\int_x^\infty yf(y)dy = x(1\text{-}F(x)) + (1\text{-}F(x))$

Differentiating both sides of the above equation, we obtain

$-xf(x) = -xf(x) + 1\text{-}F(x) -- f(x).$

On simplification, we obtain

$\dfrac{f(x)}{1-F(x)} = 1$

The solution of the equation will give the distribution of X as the exponential distribution.

The following Theorem is based on the given largest order statistics.

Theorem 2.1.5.

Suppose that X_i, i = 1,2,...,n are independent and identically distributed random variables with cdf $F(x)$ and pdf $f(x)$. We assume that $E(x)$ exists. Then $E(\overline{X} \,|X_{n,n} = x) = x + \dfrac{n-1}{n} - \dfrac{n-1}{n}\dfrac{x}{1-e^{-x}}$, and $\overline{X} = \frac{1}{n}(X_1 + X_2 + .. + X_n)$ iff

$F(x) = 1\text{-}e^{-x}$, $x \geq 0.$

Proof

Suppose $F(X) = 1\text{-}e^{-x}$.
We have

$E(\overline{X} \,|X_{n,n} \text{-}x) = E(X_1|X_{n,n} = x) = \dfrac{x}{n} + \dfrac{n-1}{n}\dfrac{\int_0^x uf(u)du}{F(x)}$

$= \dfrac{x}{n} + \dfrac{n-1}{n}\dfrac{\int_0^x ue^{-u}du}{1-e^{-x}}$

$$= \frac{x}{n} + \frac{n-1}{n} \frac{1-e^{-x}-xe^{-x}}{1-e^{-x}}$$

$$= x + \frac{n-1}{n} \left(\frac{1-e^{-x}-xe^{-x}}{1-e^{-x}} - x\right)$$

$$= x + \frac{n-1}{n} - \frac{n-1}{n} \frac{x}{1-e^{-x}})$$

$$=$$

Suppose that

$$E(X_1|X_{n,n} = x + \frac{n-1}{n} - \frac{n-1}{n} \frac{x}{1-e^{-x}}$$

Thus

$$\frac{x}{n} + \frac{n-1}{n} \frac{\int_0^x u f(u)du}{F(x)} = x + \frac{n-1}{n} - \frac{n-1}{n} \frac{x}{1-e^{-x}}$$

i.e.:

$$\int_0^x u f(u)du = F(x)\left(\frac{1-e^{-x}-xe^{-x}}{1-e^{-x}}\right)$$

Differentiating both sides of the above equation with respect to x, we get

$$xf(x) = f(x)\left(\frac{1-e^{-x}-xe^{-x}}{1-e^{-x}}\right) + F(x)\left(\frac{e^{-x}}{(1-e^{-x})^2}\right)(x + e^{-x} - 1)$$

$$\frac{x + e^{-x}-1}{1-e^{-x}}f(x) = F(x)\left(\frac{e^{-x}}{(1-e^{-x})^2}\right)(x + e^{-x} - 1)$$

On simplification

$$\frac{f(x)}{F(x)} = \frac{e^{-x}}{1-e^{-x}}$$

On integrating the both sides of the equation and using the conditions $F(0) = 0$ and $F(\infty) = 1$, we obtain

$F(x) = 1 - e^{-x}$.

However, it can be shown that if the cdf is $F(x) = 1 - e^{-x}, x \geq 0$, then

$$E(\bar{X} | X_{r,n} = x) = x + \frac{n-1}{n} - \frac{r-1}{n} \frac{x}{1-e^{-x}}).$$

We have seen that for $r = 1$ and $r = n$, the above relation characterizes the exponential distribution. It is not known whether for any other r not equal to 1 and n will characterize the exponential distribution.

Hamedani et al. [2008] gave the following characterization theorem.

Theorem 2.1.6.

Let X be a non-negative continuous random variable with cdf $F(x)$ such that $\lim\limits_{x \to \infty} x(1 - F(x))^n = 0$. Then X has an exponential distribution with parameter $\lambda > 0$ iff

$$E(X_{1,n} | X_{1,n} > t) = \frac{n\lambda + t}{n\lambda}.$$

Bairamov et.al. (2002) each of the following properties is a characteristic property of the exponential distribution.

$$P(X_{n,n} > t + x | X_{1,n} > t) = P(X_{n-1,n-1} > x)$$
$$E(X_{n,n} - t | X_{1,n} > t) = E(X_{n-1,n-1}).$$

The following theorem gives a characterization using higher moment of the random variable.

Theorem 2.1.7.

Suppose that X is an absolutely continuous random variable with cdf F(x) and pdf f(x). We assume $F(0) = 0$ and $F(x) > 0$ for all $x > 0$ and $E(X^m)$, where m is a positive integer, is finite then

$$E(X^m_{r+1,n}|X_{r,n}=x)=x^m + \frac{m}{n-r}x^{m-1} + \frac{m(m-1)}{(n-r)^2}x^{m-2} +$$

$$.... + \frac{m!}{(n-r)^{m-1}}x + \frac{m!}{((n-r)^m}, 1\leq r < n, m \geq 1.$$

if and only if $F(x) = 1 - e^{-x}, x \geq 0$.

Proof

For an alternative proof of the Theorem see Afify [2013].

$$E(X^m_{r+1,n}|X_{r,n} = x) = \int_x^\infty x^m(n-r)(\frac{1-F(y)}{1-F(x)})^{n-r-1}\frac{f(y)}{1-F(x)}dy$$

If $F(x) = 1 - e^{-x}$, then

$$E(X^m_{r+1,n}|X_{r,n} = x) = \int_0^\infty (x+u)^m(n-r)e^{-(n-r)u}du$$

$$= \int_0^\infty (x+\frac{w}{n-r})^m e^{-w}dw$$

$$= \sum_{k=0}^m \frac{m!}{k!(m-k)!}\int_0^\infty x^{m-k}(\frac{w}{n-r})^k e^{-w}dw$$

$$= \sum_{k=0}^m \frac{m!}{(m-k)!}\frac{x^{m-k}}{(n-r)^k}$$

$$= x^m + \frac{m}{n-r}x^{m-1} + \frac{m(m-1)}{(n-r)^2}x^{m-2} + ... + \frac{m!}{(n-r)^{m-1}}x$$

$$+ \frac{m!}{((n-r)^m}.$$

Suppose that

$$E(X^m_{r+1,n}|X_{r,n}=x)=x^m + \frac{m}{n-r}x^{m-1} + \frac{m(m-1)}{(n-r)^2}x^{m-2} +$$

$$.... + \frac{m!}{(n-r)^{m-1}}x + \frac{m!}{(n-r)^{m-1}},$$

then

$$\int_x^\infty x^m (n-r)\left(\frac{1-F(y)}{1-F(x)}\right)^{n-r-1}\frac{f(y)}{1-F(x)}dy$$
$$= x^m + \frac{m}{n-r}x^{m-1} + \frac{m(m-1)}{(n-r)^2}x^{m-2} + \dots + \frac{m!}{(n-r)^{m-1}}x$$
$$+ \frac{m!}{((n-r)^m}.$$

i.e.:

$$\int_x^\infty x^m (n-r)(1-F(y))^{n-r-1}f(y)dy$$
$$=(1-F(x))^{n-r}(x^m + \frac{m}{n-r}x^{m-1} + \frac{m(m-1)}{(n-r)^2}x^{m-2} +$$
$$\dots + \frac{m!}{(n-r)^{m-1}}x$$
$$+ \frac{m!}{((n-r)^m}).$$

On differentiating both sides of the equation with respect to x , simplifying and using the boundary conditions F(0) = 0 and F(∞) = 1, we obtain,$(x) = 1 - e^{-x}$, x$\geq$ 0.

Now we consider the extended sample. Suppose in addition to n observations, we have another m ($\geq$1) observations from the same population. We order the m + n observations as well as the n observations. The following theorem was given by Ahsanullah and Nevzorov (1999).

Theorem 2.1.8.

Suppose that X is an absolutely continuous random variable with cdf F(x) and pdf f(x). We assume F(0) = 0 and F(x) > 0 for all x > 0 and E(X) $_i$is finite then E($X_{i.m}|X_{j,m+n} = x$) = x + b. iff

$F(x) = 1-e^{-x}$, x$\geq$0 and b = $\frac{m}{n(m+n)}$.

It can be shown (see Ahsanullah and Nevzorov (2000)) that

If $E(X_{1,n} - X_{1,m+n}|X_{1,m+n} = x) = \varphi(x)$, then

$F(x) = 1-\exp(b-\int_0^x \{ \varphi'(u)-\frac{m}{n+m}/n\, \varphi(u)\}du$, where b is a constant.

Let $\varphi(x) = c$, c > 0, then

$F(x) = 1-\exp(b-\frac{mx}{n(m+n)c})$.

Dolegowski and Wesolowski (2015) gave the following theorem.

Theorem 2.1.9.

Suppose that X is an absolutely continuous random variable with cdf $F(x)$, pdf f(x) and E(X) is finite, then $E(X_{i,m}\}X_{j,m+n} = x) = x + b$, j < i or j > n-m + i iff X has the exponential distribution.

2.2. CHARACTERIZATION BY EQUALITY OF DISTRIBUTIONS

It is known that for exponential distribution $nX_{1,n}$ and X are identically distributed for all n $\geq$ 1. Desu (1971) proved that if $nx_{1,n}$ and X are identically distributed for all n, then the distribution of X is exponential Desu's result was improved by Arnold (1971). He showed that instead of all n, the equality of the distribution of $nX_{1,n}$ and X for two relatively prime integers greater than one will characterize the exponential distribution. Ahsanullah (1976) proved in the following theorem that with a mild condition on F(x) the identical distribution (n-i + 1) $(X_{i,n} - X_{i-1,n})$ and $nX_{1,n}$ for a fixed n and I, $1 < i \leq n$ will characterize the exponential distribution.

Theorem 2.2.1.

Let X be a non-negative random variable having an absolutely continuous (with respect to Lebesgue measure) strictly increasing distribution function F(x) for all $x \geq 0$ and F(x) < 1 for all $0 < x < \infty$. Then the following two properties are equivalent.

(1) X has an exponential distribution with F(x) = $1-e^{-\sigma x}$, $x \geq 0, \sigma > 0$,

(2) X has a monotone hazard rate and one i and n with $2 \leq i \leq n$, the statistics (n-i + 1)$(X_{i,n} - X_{i-1,n}$ and $nX_{1,n}$ are identically distributed.

We will l call a distribution function " new better than used (NBU) if $\bar{F}(x + y) \leq \bar{F}(x)\bar{F}(y)$ for all x, y ≥ 0 and "new worse than used (NWU)" if $\bar{F}(x + y) \geq \bar{F}(x)\bar{F}(y)$ for all x, y ≥ 0. We say that F belongs to class C_2 if F is either NBU or NWU.

Ahsanullah [1977] gave the following characterization of the exponential distribution based on the equality of the distribution of X and standardized spacings of the order statistics.

Theorem 2.2.2.

Let be a non-negative random variable having an absolutely continuous cdf F(x) that is strictly increasing on $(0,\infty)$. Then the following two statements are equivalent.

(i) X has an exponential distribution with $F(x) = 1 - e^{-\lambda x}$ and, $x \geq 0$ and $\lambda > 0$.

(2) For some i and n,$1 \leq i < n$, the statistics $(n-i)$ $(X_{i+1,n} - X_{i,n})$ and X are identically distributed and X belongs to class C_2.

Proof

It is known (see Galambos (1975)) that (a) $\Rightarrow$ (b). We will prove here (b) $\Rightarrow$ (a).

We can write the pdf $f_Z(z)$ of $Z = (n-i)$ $(X_{i+1,n} - X_{i,n})$ as

$$f_Z(z) = \frac{n!}{(i-1)!(n-i)!} \int_0^\infty (F(u))^{i-1} (1 - F(u + \frac{z}{n-i})^{n-i-1} f(u)$$
$$.f(u + \frac{z}{n-i})du$$

Using the assumption $f_Z(z) = f(z)$, where $f(z)$ is the pdf of x, and writing

$$\int_0^\infty (F(u))^{i-1} (1 - F(u)^{n-i} f(u)du = \frac{(i-1)!(n-i)!}{n!}$$

we obtain

$$0 = \int_0^\infty (F(u))^{i-1} g(u,z)f(u)du, \text{for all } z \geq 0 \tag{2.2.1}$$

where $g(u, z) = f(z)(1 - F(u))^{n-i} - (1 - F(u + \frac{z}{n-i}))^{n-i-1} f(u + \frac{z}{n-i})$.

Integrating (2.2.1) with respect z from 0 to z_1, we obtain

$$0 = \int_0^\infty (F(u))^{i-1} G(u, z_1)f(u)du, \text{for all } z \geq 0 \tag{2.2.2}$$

where

$$G(u,z_1) = \left(\frac{1-F\left(u + \frac{z_1}{n-i}\right)}{1-F(u)}\right)^{n-i} - (1 - F(z_1))$$

If F is NBU, then for any integer $k > 0, 1 - F\left(\frac{x}{k}\right) \geq (1 - F(x))^{\frac{1}{k}}$ so $G(0,z_1) \geq 0$. Thus if (2.2.2) holds, then $G(0,z_1) = 0$. Similarly if F is NWU, then $G(0,z_1) \leq 0$ and hence for (4.16) to be true, we must have $G(0,z_1) = 0$. Writing $G(0,z_1)$ in terms of F, we obtain

$$\left(\frac{1-F\left(u + \frac{z_1}{n-i}\right)}{1-F(u)}\right)^{n-i} = 1 - F(z_1) \qquad (2.2.3)$$

for all z_1 and any fixed $n > i$.

The solution of the above equation (see Aczel [1966]) with the boundary conditions $F(0) = 0$, $F(x) > 0$ for all x and $F(\infty) = 1$ is $F(x) = 1 = e^{-\lambda x}$, $\lambda > 0$ and $x \geq 0$.

The following theorem (Ahsanullah (1976)) gives a characterization of the exponential distribution based on the equality of two spacings.

Theorem 2.2.3.

Let X be a non-negative random variable with an absolutely continuous cumulative distribution function F(x) that is strictly increasing in $[0,,\infty)$ and having probability density function f(x). Then the following two conditions are identical.

(a)   F(x) has an exponential distribution with
$$F(x) = 1 - e^{-\lambda x}, x \geq 0$$

(b)   for some i,j and $0 \leq i < j < n$ the statistics $D_{j,n}$ and $D_{i,n}$ are identically distributed and if F has monotone increasing or decreasing hazard rate.

Proof

We have already seen (a) $\Rightarrow$ (*b*). We will give here the proof of
(b) $\Rightarrow$ (a).

The conditional pdf of $D_{j,n}$ given $X_{i,n} = x$ is given by

$$
f_{D_{i,n}}(d|X_{i,n} = x)
$$

$$
= k \int_0^\infty \left(\frac{\bar{F}(x) - \bar{F}(x+s)}{\bar{F}(x)}\right)^{j-i-1} \left(\frac{\bar{F}(x+s+\frac{d}{n-j})}{\bar{F}(x+s)}\right)^{n-j-1}
$$

$$
\cdot \frac{f(x+s)}{\bar{F}(x)} \frac{f(x+s+\frac{d}{n-j})}{\bar{F}(x+s)} \tag{2.2.4}
$$

where $k = \dfrac{(n-i)!}{(j-i-1)!(n-j)!}$

Integrating the above equation with respect to d from d to ∞ we
obtain

$$
\bar{F}_{D_{i,n}}(d|X_{i,n} = x)
$$

$$
= k \int_0^\infty \left(\frac{\bar{F}(x) - \bar{F}(x+s)}{\bar{F}(x)}\right)^{j-i-1} \left(\frac{\bar{F}(x+s+\frac{d}{n-j})}{\bar{F}(x+s)}\right)^{n-j} \frac{f(x+s)}{\bar{F}(x)}
$$

$$
\frac{f(x+s+\frac{d}{n-j})}{\bar{F}(x+s)} \, ds \tag{2.2.5}
$$

The conditional probability density function f $_{i.n}$ of $D_{i,n}$ given
$X_{i,n} = x$ is given by

$$
f_{D_{i+1,n}}(d|X_{i,n} = x) = (n-i)\left(\frac{\bar{F}\left(x+\frac{d}{n-r}\right)}{\bar{F}(x)}\right)^{n-i-1} \frac{f\left(x+\frac{d}{n-i}\right)}{\bar{F}(x)}
$$

The corresponding cdf $F_{D_{i+n}}$ is giving by

$$1 - F_{D_{i+n}}(d) = \left(\frac{\bar{F}\left(x + \frac{d}{n-i}\right)}{\bar{F}(x)}\right)^{n-i}$$

Using the relations

$$\frac{1}{k} = \int_0^{\infty} \left(\frac{\bar{F}(x+s)}{\bar{F}(x)}\right)^{n-j} \left(\frac{F(x)-F(x+s)}{\bar{F}(x)}\right)^{j-i-1} \frac{f(x+s)}{f(x)} ds$$

and the equality of the distribution of $D_{i,n}$ and $D_{j,n}$ given $X_{i,n}$, we obtain

$$\int_0^{\infty} \left(\frac{\bar{F}(x+s)}{\bar{F}(x)}\right)^{n-j} \left(\frac{F(x)-\bar{F}(x+s)}{\bar{F}(x)}\right)^{j-i-1} G(x,s,d)\frac{f(x+s)}{f(x)} ds = 0 \qquad (2.2.6)$$

where

$$G(x,s,d) = \left(\frac{\bar{F}\left(x + \frac{d}{n-i}\right)}{\bar{F}(x)}\right)^{n-i} - \left(\frac{\bar{F}\left(x + s + \frac{d}{n-j}\right)}{\bar{F}(x+s)}\right)^{n-j} \qquad (2.2.7)$$

Differentiating (2.2.7) with respect to s, we obtain

$$\frac{\partial}{\partial x} G(x,s,d) = \left(\frac{\bar{F}\left(x+s+\frac{d}{n-j}\right)}{\bar{F}(x+s)}\right)^{n-j} \left(r(x+s+\frac{d}{n-i})-r(x+s)\right) \qquad (2.2.8)$$

a) If F has increasing hazard rate , then $G(x,s,d)$ is increasing with s. Thus (2.2.7) to be true, we must have $G(x,0,d)$ $\leq G(x,s,d) \leq 0$. If F has IFR, then $\ln \bar{F}$ is concave and

b) $(\bar{F}(x + \frac{d}{n-i}))^{n-i} \geq (\bar{F}(x))^{j-i}(\bar{F}(x + \frac{d}{n-j})^{n-j}$

$\ln\bar{F}\left(x + \frac{d}{n-i}\right) \geq \frac{j-i}{n-i}\ln\bar{F}(x) + \frac{n-j}{n-i}\ln\bar{F}(x + \frac{d}{n-j})$

i.e.:

$$(\bar{F}(x + \tfrac{d}{n-i}))^{n-i} \geq (\bar{F}(x))^{j-i}(\bar{F}(x + \tfrac{d}{n-j})^{n-j}$$

Thus $G(x,0,d) \geq 0$. Thus (2.2.8) to be true we must have $G(x,0,d) = 0$ for all d and any given x. ·

(ii) If F has decreasing hazard rate, then similarly we get $G(x,0,d) = 0$. Taking x = o, we obtain from $G(x,0,d)$ as

$$(\bar{F}(\tfrac{d}{n-i})^{n-i} \ = (\bar{F}(\tfrac{d}{n-j}))^{n-j} \tag{2.2.9}$$

for all $d \geq 0$ and some i,j,n with $1 \leq i < j < n$.

Using $\varphi(d) = \ln(\bar{F}(d))$ we obtain

$$(n-i)\ \varphi\!\left(\frac{d}{n-i}\right) = (n-j)\varphi(\frac{d}{n-j})$$

Putting $\dfrac{d}{n-i} = t$, we obtain

$$\phi(t) = \tfrac{n-j}{n-i}\phi(\tfrac{n-i}{n-j}t)) \tag{2.2.10}$$

The non zero solution of (2.2.10) is

$$\varphi(x) \ = \ x \ . \text{ for all } x \geq 0. \tag{2.2.11}$$

Using the boundary conditions $F(x) = 0$ and $F(\infty)1$, we obtain

$F(x) = 1 - e^{-\lambda x}$, for all $x \geq 0$ and $\lambda > 0$. $\hspace{2cm}$ (2.2.12)

Ahsanullah (1974) gave the following theorem.

Theorem 2.2.4.

Let X be random variable with absolutely continuous cdf $F(x)$. We assume $F(0) = 0$ and $F(x) > 0$ fpr all $x > 0$.A necessary and sufficient condition that $f(x)$ will be exponential with $F(x) = 1-\exp(-\lambda x)$ is that for any fixed r and two distinct number s_1 and s_2 with $1 \leq r < s_1 < s_2 \leq n$, the distribution of the statistics $X_{s_i} - X_{r,n}$ and $X_{s_i - r. n - r}$, $i = 1,2$ are identical.

Under a mild condition on $F(x)$ Ahsanullah (1984) gave a characterization of the exponential distribution by considering one value of s instead of two values of s. We assume $F(x)$ belongs to the class c_1 if the hazard rate $r(x)$ $(= \frac{f(x)}{1 - F(x)})$ is monotone increasing or decreasing.

Theorem 2.2.5.

Let X be a non-negative random variable with an absolutely continuous cdf $F(x)$, pdf $f(x)$ with $F(0) = 0$ and $F(x) < 1$ for all $x > 0$.Then the following properties are equivalent:

(1) X has the cdf $F(x) = 1 - \exp(-\lambda x)$,
(2) $X_{k+r,n} - X_{k,n}$, for $1 \leq r < n$, $1 \leq < k < n-r$, and $X_{r,n-k}$ are identically distributed and F belongs to the class c_1 .

Proof
$\hspace{1cm}$ (a) $\Longrightarrow$ (b).
$\hspace{1cm}$ The pdf $f_1(v)$ of $X_{k+r,n} - X_{k,n}$ as

$$f_1(v) = \frac{n!}{(k-1)!(r-1)!(n-k-r)!} \int_0^\infty (F(u))^{k-1} (F(u+v) -$$

$$F(u))^{r-1} (1 - F(u+v))^{n-k-r} f(u)f(u+v)du, 0 < v < \infty.$$

$$(2.2.13)$$

$= 0$, otherwise.

The pdf $f_2(v)$ Of $X_{r,n-k}$ can be written as

$$f_2(v) = \frac{(n-k)!}{(r-1)!(n-k-r)!} (F(v))^{r-1} (1 - F(v))^{n-k-r} f(v). \qquad (2.2.14)$$

Substituting $F(x) = 1 - \exp(-\lambda x)$ in (2.2.13) and (2.2.14), the result follows.

Proof

(b) $\Longrightarrow$(a).

Integrating with respect to v from v_0 to infinity, we obtain from (2.2.13) and (2.2.14)

$$1-F_1(v_0) = \frac{n!}{(k-1)!(r-1)!(n-k-1)!} \int_0^\infty \sum_{i=0}^{r-1}(-1)^i \binom{r-1}{i} \frac{h(u,v_0)}{n-k-r+i+1} du$$

$$(2.2.15)$$

where

$$h(u,v_0)) = (F(u))^{k-1} (1 - F(u))^{r-1-i} (1 - F( ))^{n-k-r+1} f(u)$$

$$1-F_2(v_0) = \frac{(n-k)!}{(r-1)!(n-k-r)!} \sum_{i=0}^{r-1}(-1)^i \binom{r-1}{i} \frac{(1-F(v_0))^{n-k-r+i+1}}{n-k-r+i+1})$$

$$(2.2.16)$$

Since $\frac{(n-k)!}{n!(k-1)!} = \int_0^\infty (F(u))^{k-1} (1 - F(u))^{r-1-i} f(u) \, du$, using (2.2.15) and (2.2.16), we can write

$$0 = \int_0^\infty (F(u))^{k-1} (1 - F(u))^{n-k} g(u, v_0) du, \text{ for all } v_0 \geq 0,$$

$$(2.2.17)$$

where

$$g(u, v_0) = \sum_{i=0}^{r-1} (-1)^i \binom{r-1}{i} \frac{1}{n-k-r+i+1} ((1 -$$

$$F(v_0))^{n-k-r+i+1} - (\frac{1-F(u+v_0)}{1-F(u)})^{n-k-r+i+1}).$$

But $\frac{\partial g(u,v_0)}{\partial u} = \sum_{i=0}^{r-1} (-1)^i \binom{r-1}{i} (\frac{1-F(u+v_0)}{1-F(u)})^{n-k-r+i+1} (r(u +$

$v_0) - r(u))$

$$= (\frac{1-F(u+v_0)}{1-F(u)})^{n-k-r+1} (1 - \frac{1-F(u+v_0)}{1-F(u)})^{r-1} (r(u + v_0) - r(u))$$

Thus if r(x) is monotonic increasing, then for a fixed $v_0, g(u, v_0)$ is increasing in u. Since $g(0, v_0) = 0$, therefore (2.2.19) to be true, we must have

$g(u, v_0) = 0$ for all v_0, $0 \leq v_0 < \infty$ and almost all u, $0 < u < \infty$ (2.2.18)

Differentiating (2.2.18) with respect to u, we must have

$\frac{\partial g(u,v_0)}{\partial u} = 0$ for all v_0, $0 \leq v_0 < \infty$ and almost all u, $0 < u < \infty$.

$$(2.2.19)$$

Thus we obtain from (2.2.19)

r(u + v0) = r(u) for all v_0, $0 \leq v_0 < \infty$ and almost all u, $0 < u < \infty$.

Hence X is exponentially distributed. If r(x) is monotonically decreasing, then by the same argument, it follows that X is exponentially distributed.

Kakosyan et al. (1984) conjectured that the identical distribution of $p \sum_{i=1}^{M} X_i$ and X, where $P(M = k) = p(1-p)^{k-1}$, $k = 1,2,\ldots$ characterizes the exponential distribution. Ahsanullah (1988) proved the conjecture and presented the following theorem.

Theorem 2.2.6.

Let X be independent and identically distributed with cdf F(x), F(0) = 0 and F(x) > 0 for all x > 0. We assume M is an integer with $P(M = k) = p(1-p)^{k-1}$, k = 1,2... .Then the following two properties are equivalent.

a) X's have exponential distribution with cdf F(x) 1-exp($-\lambda x$). $X \geq$ 0 and $\lambda > 0$.

b) $p \sum_{i=1}^{M} X_i$ and $D_{r,n}$, where $D_{r,n} = (n-r + 1)(X_{r,n} - X_{r-1,n}), 1 \leq r \leq n$, n ≥ 2, are identically distributed and F belongs to the class c_1 with

$$\lim_{x \to 0} \frac{1-F(x)}{x} = \lambda, \lambda > 0 \text{ and } E(X) \text{ exist.}$$

Proof

Let $r \geq 2$.

Let $\phi_1(t)$ and ϕ_2T0 be the characteristic function of $p \sum_{i=1}^{M} X_i$ and $D_{r,n}$ respectively.

$$\phi_1(t) = E(e^{itp \sum_{i=1}^{M} X_i})$$

$$= \sum_{k=1}^{\infty} p(1-p)^{k-1} (\phi(pt))^k$$

$$= p\,\phi(pt)(1 - q\,\phi(pt))^{-1}$$

where $\phi(t)$ is the characteristic function of X's and q = 1-p.

If X has exponential distribution with cdf $F(x) = 1-e^{-\lambda x}$, then

$$\phi(t)) = \frac{1}{1-\frac{it}{\lambda}} \quad \text{and} \quad \phi_1(t) = \frac{p}{1-ipt\lambda^{-1}}\left(1 - \frac{q}{1-ipt\lambda^{-1}}\right)^{-1} = \frac{1}{1-\frac{it}{\lambda}} = $$

$\phi(t)$.

$$\phi_2(t) = \int_0^\infty \int_0^\infty \frac{n!e^{itv}}{(r-2)!(n-r)!}$$

$$(F(u))^{r-2}\left(1 - F\left(u + \frac{v}{n-r+1}\right)\right)^{n-r} f(u)f(u + \frac{v}{n-r+1})dudv$$

$$= 1 \quad + \quad it \quad \int_0^\infty \int_0^\infty \frac{n!e^{itv}}{(r-2)!(n-r+1)!} \quad (F(u))^{r-2}\left(1 - F\left(u + \right.\right.$$

$$\left.\left.\frac{v}{n-r+1}\right)\right)^{n-r+1} f(u)dudv$$

Substituting $F(x) = 1-e^{-\lambda x}$ anf $f(x) = \lambda e^{-\lambda x}$ in the above equation, we obtain

$$\phi_2(t) = 1 + it \int_0^\infty \int_0^\infty \frac{n!e^{itv}}{(r-2)!(n-r+1)!} \cdot ((1 -$$

$$e^{-\lambda u})^{r-2} e^{-\lambda(n-r+1)(u+\frac{v}{n-r+1})} \lambda e^{-\lambda v} dudv". = 1$$

$$+ it\int_0^\infty e^{itv} e^{-\lambda v} dv = 1 + it\frac{1}{\lambda-it} it = \frac{1}{1-\frac{t}{\lambda}} = \phi_1(t).$$

Thus (a)$\Longrightarrow$ (b).

We now prove (b)$\Longrightarrow$ (a).

Assuming $\sum_{i=1}^M X_i$ and D_r,as identically distributed and using their characteristic function, we can write as

$$P\,\phi(pt)((1 - q\,\phi(pt))^{-1}1+ it \int_0^\infty \int_0^\infty \frac{n!e^{itv}}{(r-2)!(n-r+1)!} (F(u))^{r-2}$$

$$\left(1 - F\left(u + \frac{v}{n-r+1}\right)\right)^{n-r+1} f(u)dudv$$

i.e.:

$$\frac{\phi(pt)-1}{(1-q\,\phi(pt)}\quad\frac{1}{it} = \int_0^\infty \int_0^\infty \frac{n!e^{itv}}{(r-2)!(n-r+1)!}\quad (F(u))^{r-2}\left(1 - F\left(u + \frac{v}{n-r+1}\right)\right)^{n-r+1} f(u)dudv$$

Taking limit of both sides of the equation as t goes to zero, we obtain

$$\frac{\phi'(0)}{i} =$$

$$\int_0^\infty \int_0^\infty \frac{n!}{(r-2)!(n-r+1)!}(F(u))^{r-2}\left(1 - F\left(u + \frac{v}{n-r+1}\right)\right)^{n-r+1} f(u)dudv.$$

Writing $\frac{\phi'(0)}{i} = \int_0^\infty (1 - F(v))dv$, we obtain from the above equation

$$\int_0^\infty \int_0^\infty \frac{n!}{(r-2)!(n-r+1)!}(F(u))^{r-2}(1 - F(u))^{n-r=1}$$

$$H(u,v)f(u)dudv = 0 \qquad\qquad (2.2.20)$$

where $H(u,v) = \left(\frac{1-F(u + \frac{v}{n-r+1)})}{1-F(u)}\right)^{n-r+1} - (1 - F(v))$.

It is proved by Ahsanullah (1988) that if F belongs to class c_1, then

$$H(0,v) = o \text{ for all } v \geq 0.$$

Thus

$$(1 - F\left(\frac{v}{n-r+1}\right))^{n-r+1} = 1 - F(v) \tag{2.2.21}$$

For all $v \leq 0$.

Since $\lim_{x \to 0} \frac{1-F(x)}{x} = \lambda$, it follows from (2.2.21) that

$F(x) = 1 - e^{-\lambda x}$, $x > 0$ and $\lambda > 0$.

Ahsanullah et al. (2011) proved that if

$$X_{k,n} \overset{d}{=} X_{k-1, n=1} + \frac{W}{n}, \text{ for some k. } 1 \leq k \leq, n$$

where W is distributed as exponential with $F(w) = 1-\exp(-w)$, then the random variable X has the exponential distribution with cdf $F(x) = 1-\exp(-x)$, $x > 0$. In that paper they also proved that the condition $X_{n-k,n} + \frac{W}{n+1} \overset{d}{=} X_{n-k+1,n+1}$ characterizes the exponential distribution. Wesolowski and Ahsanullah (2004) proved the following two characterization results of the exponential distribution.

It is assumed that the cdf $F(x)$ is absolutely continuous (with respect to Lebesgue measure) with $F(0) = 0$ and $F(x) > 0$ for all x, $x > 0$. Then the following results are true.

Result 2.2.1.

If $X_{k+1,n} \overset{d}{=} X_{k,n} + \frac{W}{n-k}$, for some k. $1 \leq k < n-1$

where W is distributed as exponential with $F(w) = 1-\exp(-w)$, then X has the exponential distribution with cdf $F(x) = 1-\exp(-(x-\mu))$, $x > \mu$.

Result 2.2.2.

If $X_{k,n} + \frac{W_1}{n+1} \overset{d}{=} X_{k,\,n+1} + \frac{W_2}{n-k+1}$, for some k. $1 \leq k \leq n$ where W_1 and W_2 are independently distributed as exponential with F(w) = 1-exp(-w), then X has the exponential distribution with cdf F(x) = 1-exp(-x), x > 0.

Arnold and Villasenor (2013) proved the following theorem.

Theorem 2.2.7.

If the pdf f(x) of the random variable X has derivatives of all orders in the neighborhood of 0 and $X_1 + \frac{1}{2}X_2$ and max(X_1,X_2) are identically distributed, then cdf F(x) of X is F(x) = $1-e^{-x}$, x_>0, $\lambda > 0$.

They conjectured that the followings are characteristic properties of the exponential distribution.

Identical distribution of $X_{3,3}$ and $X_2 + \frac{1}{2}X_2 + \frac{1}{3}X_3$

and

Identical distribution of $X_{3,3}$ and $X_{2,2} + \frac{1}{3}X_3$.

Yanev and Chakraborty (2013) proved that the above conjectures are true

Ahsanullah and Anis (2017) proved the following theorem.

Theorem 2.2.8.

Suppose X, X_1 and X_2 are independent and identically distributed random variables with cumulative distribution function (cdf) $F(x)$ with $F(0) = 0$ and $F(0) > 0$ for all $x > 0$. Assume that $F(x)$ is absolutely continuous (with respect to Lebesgue measure) and infinitely differentiable with $\frac{dF(x)}{dx} = f(x)$ and $f(0) > 0$. Let $Z = \max(X_1, X_2)$ and $W = \min(X_1, X_2)$. Then $Z \overset{d}{=} W + X$, where $\overset{d}{=}$ denotes equality in distribution and X is independent of W, if and only if $F(x) = 1 - e^{-\lambda x}, x \geq 0$ and λ is an arbitrary positive real number.

We define the hazard rate $h(x) = \frac{f(x)}{1-F(x)}$ for $0 < F(x) < 1$. We shall say that $F(x) \in \mathcal{C}$ if $h(x)$ is either non-decreasing or non-increasing with respect to x.

The following theorem see Ahsanullah and Anis [2017] characterizes the exponential distribution using the distributional relation between the maximum and minimum of n (> 2) random variables.

Theorem 2.2.9.

Suppose $X_1, X_2, \cdots, X_n$ are n independent and identically distributed random variables with absolutely continuous (with respect to Lebesgue measure) cumulative distribution function (cdf) $F(x)$ with $F(0) = 0$ and $F(x) > 0$ for all $x > 0$. Let the corresponding probability density function (pdf) be denoted by $f(x)$. Let $Z_n = \max(X_1, X_2, \cdots, X_n), W_n = \min(X_1, X_2, \cdots, X_n)$, and $X_{i,n}, i = 1, 2, \cdots, n$ be the ith order statistic. Then the following two conditions are identical:

a) $F(x) = 1 - e^{-\lambda x}, x \geq 0, \lambda > 0$;
b) $Z_n - W_n \overset{d}{=} X_{n-1, n-1}$ and $F(x) \in \mathcal{C}$.

Let D be the class of distribution such that the cdf F(0) = 0 and the pdf f(x) allows expansion in Maclaurin series for all x > 0. The following theorems are proved by Mlosevic and Oabradovic (2014).

Theorem 2.2.10.

Let X_1, X_2,...,X_n be random sample from a distribution that belongs to class D, then $X_{k-1,n-1} + \frac{1}{n}X_n$ and X_n are identically distributed if and only if $F(x) = 1 = e^{-\lambda x}$, $\lambda > 0$ and $x \geq 0$.

Theorem 2.2.11.

Let X_1, X_2, ..., X_n be random sample from a distribution that belongs to class D, then for $1 \leq k \leq n$, if

$\frac{1}{n}X_1 + \frac{1}{n-1}X_2 + ,,, + \frac{1}{n-k+1}X_k$ and X_n are identically distributed if and only if $F(x) = 1 = e^{-\lambda x}$, $\lambda > 0$ and $x \geq 0$.

2.3. CHARACTERIZATION BY INDEPENDENCE PROPERTIES

Fisz [1958] gave the following characterization theorem of the exponential distribution.

Theorem 2.3.1.

If X_1 and X_2 are independent and identically with an absolutely continuous cdf $F(x)$.an Then $X_{2,2} - X_{1,2}$ is independent of $X_{1,2}$ iff $F(x) = 1\text{-}\exp(-\lambda x)$, $x \geq 0$ and any $\lambda > 0$.

Proof

Then if condition is easy to prove. We will prove the only if condition.

$$P(X_{2.2.}-X_{1,2} > y \,|X_{1,2} = x) = P(X_{2,2} > x + y|X_{1,2} = x) = \frac{1-F(x+y)}{1-F(x)}.$$

Since $X_{2,2} - X_{1,2}$ is independent of $X_{1,2}$, we must have
$\frac{1-F(x+y)}{1-F(x)} = g(y)$, where $g(y)$ is a function of y only. Taking $x \to 0$,

we will have

$$1-F(x + y) = (1-F(x))(1-F(y)) \qquad (2.3.1)$$

for all x, y ≥ 0.

The solution of the equation (2.3.1) with the boundary condition

$F(0) = 0$ and $F(\infty) = 1$ is

$$F(x) = = 1-\exp(-\lambda x), x \geq 0 \text{ and any } \lambda > 0. \qquad (2.3.2)$$

Lee et.al. (2012) gave the following characterization theorem.

Theorem 2.3.2.

Let X_i (i=1,2,...,n) be i.i.d random variables with common continuous distribution function F(x). Then $X_{n,n} - X_{n-1,n}$ and $X_{n-1,n}$ are independent if and only if $F(x) = 1-e^{-\lambda x}$, $\lambda > 0$, $x > 0$.

Tanis [1964] proved that among the absolutely continuous distribution the exponential distribution has the property that

$$\sum_{j=1}^{n}(X_{j,n} - X_{1,n}) \text{ and } X_{1,n} \text{ are independent.}$$

Rogers (1963) proved that if F(X) is absolutely continuous then if for m = 2,...,n-1, $X_{m+1,n} - X_{m,n}$ and $X_{m,n}$ are independent, then F is exponential. Govindarajulu (1966, 1978) proved that if F(x) is absolutely continuous, and if $X_{t,n} - X_{s,n}$ (t > s) and $X_{s,n}$ are independent, then F is exponential.

Rossberg (1972) gave the following theorem.

Theorem 2.3.3.

If the random variable has a continuous distribution, then $X_{i,n}$ and $\sum_{i=k}^{m} c_i X_{i,n}$, $1 \leq i < k < m \leq n$, where $\sum_{i=k}^{m} c_i = 0$ for , $c_k \neq 0$, for all k, characterizes the exponential distribution.

Chapter 3

CHARACTERIZATIONS
BY RECORD VALUES

We will give several characterization theorems of the exponential distribution based on record values under various assumptions.

3.1. CHARACTERIZATIONS BY INDEPENDENCE PROPERTIES

We have already seen that $X(n) - X(m)$ and $X(m)$, $n > m \geq 1$, are independent. This is a characteristic property of the exponential distribution. For $n = 2$ Tata (1969) proved the following characterization theorem.

Theorem 3.1.1.

Let $\{ X_n, n \geq 1 \}$ be an i.i.d. sequence of non- negative random variables with cdf $F(0) = 0$ and pdf $f(x)$. Then X_n has the cdf $F(x) = 1 - e^{-\lambda x}, X \geq 0, \lambda > 0$ if and only if $X(2) - X(1)$ and $X(1)$ are independent.

Proof

We know that the if condition is true. The property of the independence of $X(2) - X(1)$ and $X(1)$ will lead to the functional equation

$$\bar{F}(0)\bar{F}(x + y) = \bar{F}(x)\bar{F}(y), 0 < x,y < \infty$$

$$(3.1.1)$$

The continuous solution of this functional equation is $F(x) = 1 - e^{-\lambda x}$, $X \geq 0, \lambda > 0$. The following theorem is a generalization of theorem 3.1.1.

Theorem 3.1.2.

Let $\{ X_n, n \geq 1 \}$ be a sequence of i.i.d. random variables with common cdf $F(x)$ which is absolutely continuous and pdf $f(x)$. Assume $F(0) = 0$ and $F(\infty) = 1$. Then for $X_n \in E(0, \sigma)$ it is necessary and sufficient that $X(n)$ and $X(n + 1) - X(n)$, $n \geq 1$, are independent.

Proof

Let $Z_{n + 1,n} = X(n + 1) - X(n)$. It is easy to establish that if $X_n \in E(0, \sigma)$, then $Z_{n + 1,n}$ and $X(n)$ are independent. We will prove the sufficient condition. Suppose that $X_{(n)}$ and $Z_{n + 1,n}$ are independent. Now the joint pdf $f(z,u)$ of $Z_{n + 1,n}$ and $X(n)$ can be written as

$$f(z,u) = \frac{(R(u))^{n-1}}{\Gamma(n)} r(u) f(u+z) , 0 < u, z < \infty.$$

$$(3.1.2)$$

$$= 0, \text{ otherwise.}$$

But the pdf $f_n(u)$ of $X(n)$ can be written as

$$f(u) = \frac{(R(u))^{n-1}}{\Gamma(n)} f(u), 0 < u < \infty,$$

$$= 0, \text{ otherwise.} \tag{3.1.3}$$

Since $Z_{n+1,n}$ and $X(n)$ are independent, we get from (3.1.2) and (3.13)

$$\frac{f(u+z)}{\overline{F}(u)} = g(z), \tag{3.1.4}$$

where $g(z)$ is independent of u. Integrating (3.1. 4) with respect z from z_1 to ∞ , we get

$$\overline{F}(u) - \overline{F}(u+z_1) = \overline{F}(u)G(z_1) \tag{3.1.5}$$

where $G(z_1) = \int_0^{z_1} g(w)dw$ Now $u \to 0^+$ we get from (3.1.5)

$$\overline{F}(u+z_1) = \overline{F}(u)\,\overline{F}(z_1). \tag{3.1.6}$$

The only continuous solution of (3.1.6) with the boundary condition $F(0) = 0$ and $F(\infty) - 1$ is

$$\overline{F}(x) = e^{-\sigma x}, x \geq 0, \tag{3.1.7}$$

where σ is an arbitrary positive real number.

The following theorem is a generalization of the theorem (3.1.2)

Theorem 3.1.3.

Let $\{ X_n , n \geq 1\}$ be independent and identically distributed with common distribution function F which is absolutely continuous with F(0)

$= 0$ and $F(x) < 1$ for all $x > 0$. Then for $X_n \in E(0,\sigma)$, it is necessary and sufficient that $Z_{n,m}$ and $X(m)$ $(n > m > 0)$ are independent. Here $Z_{n,m} = X(n)-X(m)$.

Proof

The necessary condition is easy to establish. To prove the sufficient condition, we need the following lemma.

Lemma 3.1.1.

Let $F(x)$ be an absolutely continuous function and $\bar{F}(x) > 0$, for all $x > 0$. Suppose that $\bar{F}(u+v)(\bar{F}(v))^{-1} = \exp\{-q(u,v)\}$ and $h(u,v) = \{q(u,v)\}^r \exp\{-q(u,v)\} \dfrac{\partial}{\partial u} q(u,v)$, for $r \geq 0$, $h(u,v) \neq 0$, $\dfrac{\partial}{\partial u} q(u,v) \neq 0$ for any positive u and v. If $h(u,v)$ is independent of v, then $q(u,v)$ is a function of u only. Here $\bar{F}(x) = 1 - F(x)$.

Proof

Let

$$g(u) = h(u,v) = (q(u,v))^r \exp(-q(u,v)) \frac{\partial}{\partial v} q(u,v)$$

$$= \sum_{j=0}^{\infty} \frac{(-1)^j}{\Gamma(j+1)} \{q(u,v)\}^{r+j} \frac{\partial}{\partial u} q(u,v)$$

$$= \sum_{j=0}^{\infty} \frac{(-1)^j}{\Gamma(j+1)(r+j+1)} \frac{\partial}{\partial u} (q(u,v))^{r+j+1}$$

Hence

$$\sum_{j=0}^{\infty} \frac{(-1)^j}{\Gamma(j+1)} (q(u,v))^{r+j+1} \frac{1}{(r+j+1)} = c + \int g(u)du \ ,$$

$$= g_1(u) \text{ , say .} \tag{3.1.8}$$

Here $g_1(u)$ is a function of u only and c is independent of u but may depend on v.

Now letting $u \to 0^+$, we see that $q(u,v) > 0$ and hence from (3.1.8), we have c as independent of v.

Therefore

$$0 = \frac{\partial}{\partial v} \, g_1(v) \;\; = \;\; \Sigma_{j=0}^{\infty} \frac{(-1)^j}{\Gamma(j+1)} \, \{q(u,v)\}^{r+j} \, \frac{\partial}{\partial v} \, q(u,v)$$

$$= g(u) \, \frac{\partial}{\partial v} \, q(u,v) (\frac{\partial}{\partial u} \, q(u,v))^{-1}.$$

Now we know $g(u) = h(u,v) \neq 0$ and $\frac{\partial}{\partial u} \, q(u,v) \neq 0$, so we must have

$$\frac{\partial}{\partial v} \, q(u,v) = 0.$$

Thus q(u,) is a function of u.

Now we proof the sufficiency condition of the theorem 3.1.3.

The joint pdf of $Z_{n,m}$ and $X(m)$ is

$$f(z,u) = \frac{R^{m-1}(x) \, r(x)}{\Gamma(m) \, \Gamma(n-m)} \, \{R(z+x) - R(x)\}^{n-m-1} \, f(z+x)$$

for $0 < z < \infty$, $0 < x < \infty$.

The conditional pdf of $Z_{n,m}$ given $X(m) = x$ is

$$f(z|X_{(m)} = x) = \frac{1}{\Gamma(n-m)} \{R(z+x) - R(x)\}^{n-m-1} \frac{f(z+x)}{\overline{F}(x)} \tag{3.1.9}$$

for $0 < z < \infty$, $0 < x < \infty$.

Since $Z_{n,m}$ and $X_{(m)}$ are independent, we will have for all $z > 0$,

$$(R(z+x) - R(x))^{n-m-1} \frac{f(z+x)}{\overline{F}(x)} \qquad (3.1.10)$$

as independent of x.

Since $R(z + x)$-$R(x) = -\ln\frac{\overline{F}(z+x)}{\overline{F}(x)} = q(z, x)$, writing (3.1.10) in terms of q(z,x), we get

$$q(z,x)\}^{n-m-1} \exp\{-q(z,x)\} \frac{\partial}{\partial z} q(z,x) \qquad (3.1.11)$$

as independent of x. Hence by the lemma 3.1.1, we have

$$-\ln\{\overline{F}(z + x)(\overline{F}(x))^{-1}\} = q(z + x) = c(z), \qquad (3.1.12)$$

where c(z) is a function of z only. Thus

$$\overline{F}(z+x)(\overline{F}(x))^{-1} = c_1(z) \qquad (3.1.13)$$

where $c_1(z)$ is a function of z only.

The relation (3.1.13) is true for all $z \geq 0$ and any arbitrary fixed positive number x. The continuous solution of (3.1.13) with the boundary condition $\overline{F}(0) = 1$ and $\overline{F}(\infty) = 1$ is

$$\overline{F}(x) = e^{-\sigma x}, \qquad (3.1.14)$$

for $x \geq 0$ and σ is any arbitrary positive real number.

The assumption of absolute continuity of F(x) in the theorem 3.1.3 can be replaced by the continuity of F(x).

3.2. CHARACTERIZATIONS BY EQUALITY IN DISTRIBUTIONS

We have seen that if the sequence $\{ X_n, n \geq 1 \}$ of i.i.d. rvs are from $E(0,\sigma)$, then $X(n) \underset{=}{d}$ $Z_1 + Z_2 + ... + Z_n$, where $Z_1, Z_2, ... , Z_n$ are i.i.d $E(0,\sigma)$.

The following theorem gives a characterization of the exponential distribution using the following property.

If F is the distribution function of a non- negative random variable, we will call F is "new better than used" (NBU) if for x, y ≥ 0, $\bar{F}(x+y) \leq \bar{F}(x)\bar{F}(y)$, and F is " new worse than used" (NWU) if for x, y ≥ 0, $\bar{F}(x + y) \geq \bar{F}(x) \bar{F}(y)$. We will say F belongs to the class C_1 if either F is NBU or NWU.

Let X_n, n$\geq$ 1 be a sequence of i.i.d. random variables which has absolutely continuous distribution function F with pdf f and F(0) = 0,

Assume that F(x) < 1 for all x > 0. If X_n belongs to the class C_1 and $Z_{n + 1,n}$ has an identical distribution with X_k , k $\geq$ 1, then $X_k \in E(0,\sigma)$, k $\geq$ 1.

Proof

The pdf $g_{n(x)}$ of $Z_{n + 1,n}$ can be written as

$$g_n(x) = \int\limits_0^\infty \frac{(R(u))^{n-1}}{\Gamma(n)} r(u) f(u+ z) du , \; x \geq 0$$

= 0, otherwise.   (3.2.1)

By the assumption of the identical distribution of $Z_{n+1,n}$ and X_k, we must have

$$\int_0^\infty \{R(u)\}^{n-1} \frac{r(u)}{\Gamma(n)} f(u + z) \, du = f(z), \text{ for all } z > 0. \quad (3.2.2)$$

Substituting

$$\int_0^\infty \{R(u)\}^{n-1} f(u) \, du = \Gamma(n), \quad (3.2.3)$$

In (3.2.2) we have

$$\int_0^\infty \{R(u)\}^{n-1} \, r(u) \, f(u + z) \, du$$
$$= f(z) \int_0^\infty \{R(u)\}^{n-1} \, f(u) \, du \quad (3.2.4)$$

for all $z > 0$.
   Thus

$$\int_0^\infty \{R(u)\}^{n-1} \, f(u) \, [\, f(u + z) \, (\bar{F}(u))^{-1} - f(z)] du = 0 \quad (3.2.5)$$

for all $z > 0$.

Integrating the above expression with respect to z from z_1 to ∞, we get from (3.2.5),

$$\int_0^\infty \{R(u)\}^{n-1} \, f(u)[\bar{F}(u + z_1) \, (\bar{F}(u))^{-1} - \bar{F}(z_1)] \, du = 0 \quad (3.2.6)$$

for all $z_1 > 0$.

   If F(x) is NBU, then (3.2.6) is true if

$$\overline{F}(u+z_1)\,(\overline{F}(u))^{-1} = \overline{F}(z_1), \tag{3.2.7}$$

for all $z_1 > 0$.

The only continuous solution of (3.2.7) with the boundary conditions $\bar{F}(0) = 1$ and $\bar{F}(\infty) = 0$ is $\bar{F}(x) = e^{-\sigma x}$, where σ is an arbitrary real positive number.

Similarly, if F is NWU then (3,2.6) is true if (3.2.7) is satisfied and hence

$$X_k \in E(0,\sigma) , k \geq 1.$$

The following theorem is proved under the assumption of monotone hazard rate. We will say F belongs to the class C_2 if $r(x)$ is either monotone increasing or decreasing.

Theorem 3.2.3.

If X_k, $k \geq 1$ has an absolutely continuous distribution function F with pdf f and $F(0) = 0$. If $Z_{n+1,n}$ and $Z_{n,n-1}$, $n \geq 1$, are identically distributed and F belongs to C_2, then $X_k \in E(0,\sigma)$, $k \geq 1$.

Proof

$$P(Z_{n+1,n} > z) = \int_0^\infty \{R(u)\}^{n-1} \frac{r(u)}{\Gamma(n)} \bar{F}(u+z)du, \text{ for all } z > 0,$$

$$= 0, \text{ otherwise.}$$

Since $Z_{n+1,n}$ and $Z_{n,n-1}$ are identically distributed, we get using the above equation,

$$\int_0^\infty \{R(u)\}^n \; r(u) \bar{F}(u+z)\,du$$
$$= n \int_0^\infty \{R(u)\}^{n-1} r(u) \bar{F}(u+z)\,du \tag{3.2.8}$$

for all $z > 0$.

Substituting the identity

$$n \int_0^\infty \{R(u)\}^{n-1} \; r(u) \; \bar{F}(u \; + z) \; du$$
$$= \int_0^\infty \{R(u)\}^n \; f(u \; + z) \; du$$

in (3.2.8), we get on simplification

$$\int_0^\infty \{R(u)\}^{n-1} \; r(u) \bar{F}(u \; + \; z) \left[1 - \frac{r(u \; + z)}{r(u)}\right] \; du = 0, \quad (3.2.9)$$

for all $z > 0$.

Thus if $F \in C_2$, then (3.2.9) is true if for almost all u and any fixed z > 0,

$$r(u + z) = r(u) . \tag{3.2.10}$$

The constant hazard rate i.e., the relation (3.2.10) is a known characteristic property of the exponential distribution. Hence, we have $X_k \in E(0,\sigma)$ for all $k \geq 1$.

Theorem 3.2.4.

Let X_n, $n \geq 1$ be a sequence of independent and identically distributed non-negative random variables with absolutely cdf $F(x) = 0$ and pdf $f(x)$. If $F \in C_2$ and for some fixed n, m, $1 \leq m < n < \infty$, $Z_{n,m}$ and $X_{(n-m)}$ are identically distributed, then $X_k \in E(0,\sigma)$, $k \geq 1$.

Proof

The pdfs $f_1(x)$ of $X(n-m)$ and $f_2(x)$ of $Z_{n,m}$ can be written as

$$f_1(x) = \frac{1}{\Gamma(n-m)} (R(x))^{n-m-1} f(x), \text{ for } 0 < x < \infty, \quad (3.2.11)$$

and

$$f_2(x) = \int_0^\infty \frac{(R(u))^{m-1}}{\Gamma(m)} \frac{\{R(x+u) - R(u)\}^{n-m-1}}{\Gamma(n-m)} r(u) f(u + x) du, \quad (3.2.12)$$

for $0 < x < \infty$.

Integrating (3.2.11) and (3.2.12) with respect to x from 0 to x_0, we get

$$F_1(x_0) = 1 - g_1(x_0) \quad (3.2.13)$$

where

$$g_1(x_0) = \sum_{j=1}^{n-m} \frac{(R(x_0))^{j-1}}{\Gamma(j)} e^{-R(x_0)},$$

and

$$F_2(x_0) = 1 - G_2(x_0,) \text{ and}$$
$$G_2(x_0) = \frac{1}{\Gamma(m)} \int_0^\infty (R(u))^{m-1} g_2(x_0, u) f(u) du \quad (3.2.14)$$

where

$$g_2(x_0, u)$$

$$= \sum_{j=1}^{n-m} \frac{\{R(u+x_o)-R(u)\}^{j-1}}{\Gamma(j)} \exp\{-(R(u+x_o)-R(u))\} .$$

Now equating (3.2.13) and (3.2.14), we get

$$\int_0^\infty \frac{\{R(u)\}^{m-1}}{\Gamma(m)} f(u) [g_2(u,x_o) - g_1(x_o)] \, d u = 0, \quad (3.2.15)$$

for all $x_O \geq 0$.

Now $g_2(x_0,u) = g_1(x_1)$ and

$$\frac{d}{du}(g2(x0),u)- \ g1(x0)) = \frac{\{R(u)-R(u)\}^{n-m-1}}{\Gamma(n-m)} \exp\{-(R(u+$$
$$x_o)-R(u)\} [r(x_o) - r(u+x_o)] .$$

Thus if $F \in C_2$, then (3.2.15) is true if

$$r(u + x_O) = r(u) \qquad\qquad\qquad (3.2.16)$$

for almost all u and any fixed $x_O \geq 0$.

Hence $X_k \in E(0,\sigma)$ for all $k \geq 1$, here σ is an arbitrary positive real number.

Substituting $m = n-1$, we get $Z_{n,n-1} \underset{=}{d} X_1$ as a characteristic property of the exponential distribution.

Theorem 3.2.5.

Let $\{ X_n, n \geq 1 \}$ be a sequence of independent and identically distributed non-negative random variables with absolutely continuous cdf $F(x) = 0$ and the pdf $f(x)$. If F belongs to C_2 and for some m , m > 1,

X(m)-X(m) + U, where U is independent of X(m) and X(m-1) is distributed as X_n's ,then $X_n \in E(0, \sigma)$, for some $\sigma > 0$.

Proof

The pdf $f_m(x)$ of X(m) m ≥ 1, can be written as

$$f_m(y) = \frac{(R(y))^{m-1}}{\Gamma(m)} f(y), 0 < y < \infty,$$

$$= \frac{d}{dy}\{-\bar{F}(y) \int_0^y \frac{(R(x))^{m-2}}{\Gamma(m-1)} r(x) \, dx + \int_0^y \frac{(R(x))^{m-1}}{\Gamma(m-1)} f(x) \, dx)$$

$$(3.2.17)$$

The pdf $f_2(y)$ of X(m + 1) + U can be written as

$$f_2(y) = \int_0^y \frac{(R(x))^{m-2}}{\Gamma(m-1)} f(y-x) f(x) dy$$

$$= \frac{d}{dy}(-\frac{(R(x))^{m-2}}{\Gamma(m-1)} \bar{F}(y-x)f(x)dx + \int_0^y \frac{(R(x))^{m-2}}{\Gamma(m-1)} f(x) \, dx)$$

$$(3.2.18)$$

Equating (3.4.31) and (3.4.32), we get on simplification

$$\int_0^y \frac{(R(x))^{m-1}}{\Gamma(m-1)} f(x) H_1(x, y) \, dx = 0,$$

where

$$H_1(x, y) = \bar{F}(y-x) - \bar{F}(y) (\bar{F}(x))^{-1}, 0 <$$
$$x < y < \infty \qquad (3.2.19)$$

Since $F \in C_1$, therefore for (3.2.19) to be true , we must have

$$H_1(x,y) = 0 \tag{3.2.20}$$

for almost all x, $0 < x < y < \infty$. Now $H_1(x,y) = 0$ for almost all x, $0 < x < y < \infty$, implies

$$\overline{F}(y-x)\,\overline{F}(x) = \overline{F}(y) \tag{3.2.21}$$

for almost all x, $0 < x < y < \infty$.

The only continuous solution of (3.2.21) with the boundary conditions

$$\overline{F}(0) \;=\; 1 \text{ and } \overline{F}(\infty) \;=\; 0, we \text{ obtain}$$
$$\overline{F}(x) = e^{-\sigma x} \text{ , } x \geq 0 \tag{3.2.22}$$

where σ is an arbitrary positive number.

Remark 3.2.1.

The theorem 3.4.7 can be used to obtain the following known results of a two parameter exponential distribution ($\overline{F}(x) = \exp\{-\sigma^{-1}(x-\mu)\}$).

$$E(X(m) = \mu + m\,\sigma$$
$$Var(X(m)) = m\,\sigma^2$$
$$Cov(X(m)\,X(n) = m\,\sigma^2, m < n.$$

Theorem 3.2.6.

Let $X_1, ..., X_m,...$ be independent and identically distributed random variables with probability density function f(x), $x \geq 0$ and m is an integer valued random variable independent of X's and $P(m = k) = p(1-p)^{k-1}$, k = 1, 2, ... and $0 < p < 1$. Then the following two properties are equivalent:

(1) X's are distributed as $E(0,\sigma)$, where σ is a positive real number,

(2) $p\sum_{j=1}^{m} X_j$ and $X(n)-X(m)$, for some fixed n, $n \geq 2$,are identically distributed, $X_j \in c_2$ and $E(X_j) < \infty$.

Proof

It is easy to verify (a) $\Rightarrow$ (b). We will prove here that (b) $\Rightarrow$ (a).

Let $\phi_1(t)$ be the characteristic function of $X(n) = X(n-1)$), then

$$\varphi_1(t) = \int_0^\infty \int_0^\infty \frac{1}{\Gamma(n)} \; e^{itx} \; (R(u))^{n-1} \; r(u)f(u + x) \; du \; dx$$

$$= 1 + it \int_0^\infty \int_0^\infty \frac{1}{\Gamma(n)} \; e^{itx} \; (R(u))^{n-1} \; r(u)$$
$$.\bar{F}(u + x) \; du \; dx$$

$$(3.2.23)$$

The characteristic function $\Phi_2(t)$ of $p \sum_{j=1}^{m} X_j$ can be written as

$$\Phi_2(t) = E(e^{itp \sum_{j=1}^{\infty} X_j})$$
$$= \sum_{k=1}^{\infty} (\Phi(tp))^k p(1-p)^{k-1}$$

where $\Phi(t)$ is the characteristic function of X's.

$$= p(\Phi(t\; p)) \; (1 - q \; \Phi(p\; t))^{-1}, q = 1\text{-}p. \tag{3.2.24}$$

Equating (3.2.23) and (3.2.24), we get on simplification

$$\frac{\Phi(pt)-1}{1-q\;\Phi(pt)} \frac{1}{it} = \int_0^\infty \int_0^\infty \frac{1}{\Gamma(n)} \; e^{itx} \; (R(u))^{n-1} r(u) \; \bar{F}(u +$$
$$x) \; du \; dx \tag{3.2.25}$$

Now taking limit of both sides of (3.4.25) as t goes to zero, we have

$$\frac{\Phi'(0)}{i} = \int_0^\infty \int_0^\infty \frac{1}{\Gamma(n)} \ (R(u))^{n-1} \ r(u) \ \bar{F}(u + x) du \ dx \ .$$

(3.2.26)

Writing

$$\frac{\Phi'(0)}{i} = \int_0^\infty \bar{F}(x) \, dx \int_0^\infty \int_0^\infty (R(u))^{n-1} \ r(u) \ \{\bar{F}(u + x) -$$
$$\bar{F}(u) \ \bar{F}(x) \} du \ dx \ = \ 0$$

(3.2.27)

Since X's belongs to C_1, we must have

$$\bar{F}(u + x) \ = \ \bar{F}(x) \ \bar{F}(u)$$

(3.2.28)

for almost all x, u, $0 < u, x < \infty$.

The only continuous solution of (3.2.28) with the boundary condition $\overline{F}(0) = 1 \ and \ \overline{F}(\infty) = 0$, is

$$\bar{F}(x) = e^{-\sigma x}, \ x \geq 0,$$

(3.2.29)

where σ is an arbitrary positive real number.

3.3. CHARACTERIZATIONS BY MOMENTS
AND CONDITIONAL MOMENTS

We will proof the following characterization theorem under the assumption of the finite first moment.

Theorem 3.3.1.

Let X_n, $n \geq 1$ be a sequence of independent and identically distributed non-negative random variables with absolutely continuous distribution function $F(x)$ and the corresponding density function $f(x)$. Let $a = \inf\{x|F(x) > 0\} = 0$, $F(x) < 1$ for all $x > 0$. If F belongs to the class C_1 and $E(X_k)$, $k \geq 1$ is finite., then $X_k \in E(0,\sigma)$, if and only if for some fixed n, $n > 1$, $E(Z_n) = E(X_k)$.

Proof

If $X_k \in E(o,\sigma)$, then it can easily be seen that $E(Z_n) = E(X_k)$. Suppose that for some foxed n, $n > 1$, $E(Z_n) = E(X_k)$, then we must have

$$\int_0^\infty \int_0^\infty \frac{(R(u))^{n-1}}{\Gamma(n)} f(u) (\bar{F}(u))^{-1} \, du \, dx$$
$$= \int_0^\infty \bar{F}(u) \, du \tag{3.3.1}$$

But we know

$$\Gamma(n) = \int_0^\infty (R(u))^{n-1} f(u) \, du \tag{3.3.2}$$

Substituting (3.3.2) in (3.3.1) and simplifying, we have

$$\bar{F}(u + z) = \bar{F}(u) \bar{F}(z) \tag{3.3.3}$$

for all u, z, $0 < u,z < \infty$.

Now the continuous solution of (3.3.4) with the boundary conditions $\bar{F}(0) = 1$ and $F(\infty) = 0$, is $\bar{F}(x) = \exp(-x\sigma^{-1})$, where σ is an arbitrary real number and $x \geq 0$.

Nagaraja (1977) proved that under certain conditions on the cdf F(x), E(X(n + 1)|X(n) = x)) = x + c ,where c is a constant characterizes the exponential distribution. Ahsanullah and Wesolowski (1998) generalized the result of Nagaraja (1977). They gave the following theorem.

Theorem 3.3.2.

Let $\{X_n, n \geq 1\}$ be i.i.d. sequence of random variables with absolutely continuous cdf F(x) . Assume that for some c > 0,$E(X^{1+c})$ exists and F(0) = 0. If E(X(n + 2)|X(n) = x) = x + b, where b is constant, then F(x) = $1 - e^{-\lambda x}, \lambda > 0, x \geq 0$.

Deminska and Wesolowski (2000) generalized the result of Ahsanullah and Wesolowski (1998). They proved that under the same conditions of the theorem 3.3.2 , the condition

E(X(n + k)|X((n) = x) = x + d,

where $k \geq 1$ and d is a constant characterizes the exponential distribution.

It follows from the result of Deminska and Wesolowski (2000), see also Huan and Lee (1993) and Wu (2004) that

E(X(n)-X(m)) = (n-m)c ,$1 \leq m < n$,

where c is constant characterizes the exponential distribution.

The following theorem gives a characterization by higher moments.

Theorem 3.3.3.

Let $\{X_n, n \geq 1\}$ be i.i.d. sequence of random variables with absolutely continuous cdf F(x) and pdf f(x). . Assume that (X^m), where m is a positive integer, exists and F(0) = 0. Then

$E((X(n + 1)^m | X(n) = x) = x^m + mx^{m-1} + m(m-1) x^{m-2} + \ldots + m!x + m!$ if and only if $F(x) = 1 - e^{-x}, x \geq 0$.

Proof

If $F(x) = 1 - e^{-x}. x \geq 0$, then

$$E(X(n + 1)^m | X(n) = x) = \int_x^\infty y^m e^{-(y-x)} dy$$
$$= x^m + mx^{m-1} + m(m-1) x^{m-2} + \ldots + m! \, x + m!$$

Suppose that

$E(X(n + 1)^m | X(n) = x) = x^m + mx^{m-1} + m(m-1) x^{m-2} + \ldots + m!x + m!$, then

$$\int_x^\infty \frac{f(y)}{1-F(x)} dy = x^m + mx^{m-1} + m(m-1) x^{m-2} + \ldots + m!x + m!.$$

Therefore

$$\int_x^\infty f(y) \, dy = (1 - F(x))(x^m + mx^{m-1} + m(m-1) x^{m-2} + \ldots + m!x + m!).$$

Differentiating both sides of the above equation with respect to s, we obtain on simplification

$$\frac{f(x)}{1-F(x)} = 1.$$

Thus the cdf of the random variables $\{X_n, n \geq 1\}$ is exponential with $F(x) = 1 - e^{-x}, x \geq 0$.

Noor and Athar [2014] proved that under the condition of the Theorem 3.3.3., the relation

$$E((X(n + k) - X(n))^m | X(n) = x) = \frac{\Gamma(m + k)}{\Gamma(k)}$$

characterizes the exponential distribution having $F(x) = 1\text{-}e^{-x}$, $x \geq 0$. The following theorem uses the property of homoscedasticity but does not use NBU or NWU property.

Theorem 3.3.4.

Let x_n, $n \geq 1$ be a sequence of independent and identically distributed random variables with common distribution function F which is absolutely continuous $F(0) = 0$ and $E(X_n^2) < \infty$. Then X_k, $k \geq 1$ has the exponential distribution if and only if $\mathrm{var}(Z_n | X(n) = x) = b$ for all x, where b is a positive constant independent of x and $Z_n = X(n + 1) - X(n)$.

Proof
    The 'if' condition is easy to show.
    We will prove here the 'only if' condition.

$$b = E(Z_n^2 | X(n) = x) - [E(Z_n | X(n) = x_)]^2. \qquad (3.3.4)$$

Thus

$$E(Z_n^2 | X(n) = x)$$
$$= \int_0^\infty z^2 (\bar{F}(x))^{-1} d\bar{F}(z + x) = 2 \int_0^\infty z (\bar{F}(x))^{-1} \bar{F}(z + x) \\ dz \qquad\qquad (3.3.5)$$

and

$E(Z_n \mid X(n)) = x)$

$= \int_0^\infty (\bar{F}(x))^{-1} \; d\bar{F}(z + x) = \int_0^\infty (\bar{F}(x))^{-1} \bar{F}(z + x)dz$ (3.3.6)

Substituting $G(x) = \int_0^\infty z \; \bar{F}(z \mid x) \; dz$ and denoting $G^{(r)}(x)$ as the rth derivative of $G(x)$, we have on simplification

$$G^{(1)}(x) = \int_0^\infty \bar{F}(z + x) \; dz \;, G^{(2)}(x)$$
$$= \bar{F}(x) \; and \; G^{(3)}(x) = -f(x).$$

Writing (3.3.4) and (3.4.5) in terms of $G(x)$ and $G^{(r)}(x)$, we get from (3.3.6),

$$2G(x) \{G^{(r)}(x)\}^{-1} - \{ G^{(1)}(x) (G^{(2)}(x))^{-1}\}^2 = b$$ (3.3.7)

for all $x > 0$.

Differentiating (3.3.7) with respect to x and simplifying, we obtain

$$2G^{(3)}(x) \{G^{(2)}(x)\}^{-3} - [(G^{(1)}(x))^2 - G(x) \; G^{(2)} (x)] = 0$$ (3.3.8)

Since $G^{(3)} (x) \neq 0$ for all $x > 0$, we must have

$$\{G^{(1)}(x)\}^2 - G(x) \; G^{(2)}(x) = 0,$$ (3.3.9)

i.e.:

$$\frac{d}{dx}\{ G(x) (G^{(1)}(x))^{-1}\} = 0, \text{ for all } x > 0.$$ (3.3.10)

The solution of (3.2.10) is

$$G(x) = a\,e^{-cx}, x > 0 \tag{3.3.11}$$

where a and c are arbitrary constants.
Hence

$$\overline{F}(x) = G^{(2)}(x) = ac^2 e^{-cx}, x \geq 0.$$

Since F(x) is a distribution function with $F(0) = 0$, it follows that

$$\overline{F}(x) = e^{-\sigma x}, x \geq 0,$$

where σ is an arbitrary real positive number.

3.4. CHARACTERIZATIONS BY EQUALITY IN HAZARD RATES

The following Theorem gives a characterization of the exponential distribution using the hazard rate.

Theorem 3.4.1.

Let $\{X_n, n \geq 1\}$ be a sequence of independent and identically distributed non negative random variables with absolutely continuous distribution function F(x) and the corresponding density function f(x). Let $a = \inf\{x|F(x) = 0\} = 0$, $F(x) < 1$ for all $x > 0$ and F belongs to class C_3. Then $X_k \in E(0,\sigma)$, if and only if for some fixed n, $n \geq 1$, the hazard rate r_1 of $Z_{n+1,n}$ = the hazard rate r of X_k.

Proof

If $X_k \in E(0,\sigma)$, then it can easily be shown that $r_1 = r$.

Suppose $r_1 = r$. We can write the joint pdf of $X(n + 1)$ and $X(n)$ as

$$f_{n+1,n}(x,y) = \frac{1}{\Gamma(n)} \{R(x)\}^{n-1} r(x) f(y), \ 0 < x < y < \infty,$$

$= 0$, otherwise.

Substituting $Z_{n+1,n} = X(n+1) - X(n)$ and $V_n = X(n)$, we get the pdf of $Z_{n+1,n}$ and V_n as

$$f_1^*(z,u) = \frac{1}{\Gamma(n)} \{R(u)\}^{n-1} r(u) f(u+z) \tag{3.4.1}$$

for $0 < u, z < \infty$,

$= 0$, otherwise.

Thus by (3.4.1), we can write

$$r_1(z) = \frac{\int_0^\infty (R(u))^{n-1} r(u) f(u+z) du}{\int_0^\infty R(u))^{n-1} (r(u) \bar{F}(u+z) du} \tag{3.4.2}$$

for all $z \geq 0$. Since $r_1(z) = r(z)$ for all z, we must have

$$\frac{\int_0^\infty (R(u))^{n-1} r(u) f(u+z) du}{\int_0^\infty (R(u))^{n-1} r(u) \bar{F}(u+z) du} = \frac{f(z)}{\bar{F}(z)} \tag{3.4.3}$$

for all $z \geq 0$. Now simplifying (3.4.3) we have

$$\int_0^\infty (R(u))^{n-1} r(u) \bar{F}(z) \bar{F}(u+z) \{r(u+z) - r(z)\} du = 0 \tag{3.4.4}$$

for all $z \geq 0$. Since F belongs to class C_2, for (3.4.4) to be true, we must have

$$r(u + z) = r(u) \qquad\qquad (3.4.5)$$

for all $z > 0$ and almost all u, $u \geq 0$. Hence $X_k \in E(0,\sigma)$.

3.5. CHARACTERIZATIONS BY LOWER RECORDS

The exponential distribution can also be characterized using lower record values. Ahsanullah and Kirmani (1991) characterized the exponential distribution using lower record values. The following result is due to Ahsanullah and Kirmani [1991].

Suppose $\{X_n, n \geq 1\}$ be a sequence of i.i.d. random variables with cdf F and $F(0) = 0$. Let N is the rv defined as $N = \min \{ i > 1: X_i < X_1 \}$.

It can easily be shown that $P(N = n) = \dfrac{1}{n(n-1)}$, $n = 2,3,...$

Theorem 3.5.1.

The rvs NX_N and X_1 are identically distributed iff $F(x) = \exp(-\lambda x)$, $x > 0$, for some $\lambda > 0$.

In proving the theorem , we need the following Lemma.

Lemma 3.5.1.

$$P(NX_N > x) = \sum_{n=2}^{\infty} \frac{1}{n(n-1)} (\bar{F}(\frac{x}{n}))^n, \text{ for all } x \geq 0.$$

Proof

$$P(NX_N > x) = \sum_{n=2}^{\infty} P(NX_N > x, N > n)$$
$$= \sum_{n=2}^{\infty} \int_{-\infty}^{\infty} P(nX_n > x, X_n < y, X_i > y \text{ for all } i = 2,3,..., n-1 |$$
$$X_1 = y) \, dF(y)$$
$$= \sum_{n=2}^{\infty} \int_{\frac{x}{n}}^{\infty} P(\frac{x}{n} < x_n < y)(P(x_i > y))^{n-2} \, dF(y)$$
$$= \sum_{n=2}^{\infty} \int_{\frac{x}{n}}^{\infty} \{\bar{F}(\frac{x}{n}) - \bar{F}(y)\} \{\bar{F}(y)\}^{n-2} dF(y)$$
$$= \sum_{n=2}^{\infty} \bar{F}(\frac{x}{n}) \{ \frac{(\bar{F}(\frac{x}{n}))^{n-1}}{n-1} - \frac{(\bar{F}(\frac{x}{n}))^n}{n} \}$$
$$= \sum_{n=2}^{\infty} \frac{1}{n(n-1)} [\bar{F}(\frac{x}{n})]^n.$$

Proof of the Theorem 3.3.4.

Define

$$u(x) = - \frac{\ln \bar{F}(x)}{x}, x > 0; u(0) = u(0 +)$$

and suppose that

$$NX_N \underset{=}{d} X_1.$$

Then

$$\sum_{n=2}^{\infty} \frac{1}{n(n-1)} e^{-xu(x/n)} = e^{-xu(x)}, x > 0. \tag{3.5.1}$$

We shall show that the above holds iff u(x) is a constant, i.e. given any T > 0

$$\min_{x \in [0,T]} u(x) = \max_{x \in [0,T]} U(x). \tag{3.5.2}$$

Let

$$a_0 = \min_{x \in [0,T]} U(x), x_0 = inf\{x \in [0,T] | u(x) = a_0\},$$

$$a_1 = \max_{x \in [0,T]} U(x), x_1 = inf\{x \in [0,T] | u(x) = a_1\}.$$

It is obvious that (3.5.2) will be proved if we show that $x_0 = 0 = x_1$. By continuity of u, $x_0 \in [0,T]$ and $u(x_0) = a_0$. Hence

$$u(x_0) \leq u(x_0 / n) \text{ for all } n \geq 1. \tag{3.5.3}$$

If equality holds for all $n \geq 2$, then $u(x_0) = u(0)$ which by definition of $x_0 = 0$. Suppose now that $x_0 > 0$ (so that $x_0 / n \neq x_0$ for all n. Then, strict inequality must hold for (3.3.19) for at least one value of $n \geq 2$ and

$$e^{-x_0 u(x_0)} - \sum_{n=2}^{\infty} \frac{1}{n(n-1)} e^{-x_0 u(x_0/n)}$$

$$= \sum_{n=2}^{\infty} \frac{1}{n(n-1)} \{e^{-x_0 u(x_0)} - e^{-x_0 u(x_o/n)}\} > 0,$$

which contradicts (3.5.3). Therefore $x_0 = 0$. Similarly $x_1 = 0$. Thus

$$NX_N \underline{d} X_1 \Rightarrow u(x) \equiv \text{constant. The converse can easily be verified.}$$

Basak (1996) give a similar characterization based on k-lower records.

The result is given in the following theorem.

Theorem 3.5.2.

Suppose $\{X_n, n \geq 1\}$ be a sequence of i.i.d. random variables with cdf $F(x)$ with $F(0) = 0$.and $\lim\limits_{x \to 0^+} \dfrac{F(x)}{x} = \lambda$, $\lambda > 0$, . If $(L(n,k) - k + 1)$ $X(n,k))$ and $X_{1,n}$, $k \geq 1$, where $L(n + k)$ is the index where kth lower record occurs, are identically distributed, then X has the exponential distribution with $F(x) = 1 - \exp(-\lambda x)$, $x \geq 0$ and $\lambda > 0$.

Chapter 4

CHARACTERIZATIONS BY GENERALIZED ORDER STATISTICS

In this chapter we will give some characterizations of exponential distributions based on generalized order statistics. We know that order statistics and record values are special cases of generalized order statistics. Many characterizations of the exponential distribution based on the generalized order statistics can be obtained using the methods given in order statistics and record values.

4.1. CHARACTERIZATIONS BY CONDITIONAL EXPECTATIONS

Theorem 4.1.1.

Let X be an absolutely continuous (with respect to Lebesgue measure) non-negative random variable having cdf $F(x)$ with $F(0) = 0$, $F(x) < 1$ for all $x > 0$ and pdf $f(x)$. Then the following two properties are equivalent

a) X has an exponential distribution, E(0,1);

b) for $1 < r \le n$, the statistics $E(X(r + 1,n,m,k)| X(r,n,m,k) = x) = x$ $+ \dfrac{1}{\gamma_{r+1}}$..

Proof

The proof of (a)$\Rightarrow$ (*b*) follows from (1.4.2).

We will prove here (b)$\Rightarrow$ (a)

We have

$$E(X(r + 1,n,m,k)| X(r,n,m,k) = x) =$$

$$\int_x^\infty y\gamma_{r+1}\left(\frac{1-F(y)}{1-F(x)}\right)^{\gamma_{r+1}-1}\frac{f(y)}{1-F(x)}\,dy \qquad (4.1.1)$$

Writing $(X(r+1,n,m,k) | X(r,n,m,k) = x) = x + \dfrac{1}{\gamma_{r+1}}$, we obtain from (4.1.1)

$$\int_x^\infty y\gamma_{r+1}\left(\frac{1-F(y)}{1-F(x)}\right)^{\gamma_{r+1}-1}\frac{f(y)}{1-F(x)}\,dy = x + \frac{1}{\gamma_{r+1}}.$$

i.e.:

$$\int_x^\infty y\gamma_{r+1}(1 - F(y))^{\gamma_{r+1}-1}f(y)\,dy$$
$$= \left(x + \frac{1}{\gamma_{r+1}}\right)(1 - F(x))^{\gamma_{r+1}}$$

Differentiating both sides of the above equation with respect to x, we obtain on simplification

$$\frac{f(x)}{1-F(x)} = 1$$

Thus

$$F(x) = 1-e^{-x}, \ x \geq 0.$$

The following theorem is a generalization of Theorem 4.1.1.

Theorem 4.1.2.

Let X be a non-negative random variable having an absolutely continuous (with respect to Lebesgue measure) cdf F(x) with F(0) = 0 and F(x) < 1 for all x > 0. Then the following two properties are equivalent

a) X has an exponential distribution, E(0,1),
b) for $1 < r \leq n$, the statistics $E(X(r + 2,n,m,k)| \ X(r,n,m,k) = x) = x$
$$+ \frac{1}{\gamma_{r+1}} + . \frac{1}{\gamma_{r+2}}.$$

Proof

The proof of (a)$\Longrightarrow$ (b) follows from (1.4.2).

To proof (b) $\Longrightarrow$(a), we use the following theorem (Theorem A) given by Ahsanullah and Raqab (2004).

Theorem A

Suppose the random Variable X is absolutely continuous having cdf F(x) and pdf f(x). We assume F(0) and F(x) > 0 for all x > 0 and E(X) is finite.

$E(\emptyset(X(s,n,m,k)|X(r,n,m,k) = x) = g(x)$.then the hazard function r(x) is the solution of the equation

$$\gamma_{r+2}\gamma_{r+1}(g(x)\text{-}\emptyset(x))r^3(x)\text{-}(2\gamma_{+1}\text{-m-1})g'(x)r^2(x) + g''(x) -$$

$$\frac{r'(x)}{r(x)}g'(x) = 0.$$

Putting $\emptyset(x) = x$ and $g(x) = x + \frac{1}{\gamma_{r}+1} + \frac{1}{\gamma_{r}+1}$, in the above equation we obtain $r(x) = 1$ as the solution.

Thus $F(x) = 1- e^{-x}$, $x \geq 0$.

The following theorem is due to Ahsanullah and Hamedani [2013].

Theorem 4.1.3.

Let the random variable X have an absolutely continuous cdf $F(x)$ and pdf $f(x)$. We assume $E(X)$ exist. Then for r and r-1, $1 < s \leq n$,

$$E(X(s,n,m,k) - X(r,n,m,k)^p | X(r,n,m,k) = x)$$

$$= \frac{\Pi_{j=r+1}^{s}\gamma_j}{\lambda^p}\frac{\Gamma(p+1)}{(m+1)^{s-r-1}}[\Sigma_{j=0}^{s-r-1}\frac{\Gamma(s-r)}{\Gamma(j+1)\Gamma(s-r-j)}(-1)^j\left(\frac{1}{\gamma_{s-j}}\right)^{p+1}],$$

$$m \neq -1, p > 0,$$

$$(4.1.2)$$

If and only if $F(x) = 1-e^{-\lambda x}$, $\lambda > 0$ and $x \geq 0$.

Remark 4.1.1.

If $s = r + 2$ and $p = 1$, then (4.1.2) reduces to

$$E((X(r+2,n,m,k)-X(r,m,n,k))|X(r,n,m,k) = x)$$

$$= \frac{\Pi_{j=r+1}^{r+2}\gamma_j}{\lambda^2}\frac{\Gamma(2)}{m+1}[\Sigma_{j=0}^{1}\frac{1}{\Gamma(j+1)\Gamma(2-j)}(-1)^j\left(\frac{1}{\gamma_{r+2-j}}\right)^2]$$

$$= \frac{\gamma_{r+1}\gamma_{r+2}}{\lambda^2}\frac{1}{m+1}[\frac{1}{\gamma_{r+2}^2} - \frac{1}{\gamma_{r+1}^2}]$$

$$= \frac{\gamma_{r+1}\gamma_{r+2}}{\lambda^2}\frac{\gamma_{r+1}+\gamma_{r+2}}{\gamma_{r+2}^2\gamma_{r+1}^2}$$

$$= \frac{1}{\lambda^2}(\frac{1}{\gamma_{r+1}} + .\frac{1}{\gamma_{r+2}}).$$

Thus the Theorem 4.1.2. is a special case of Theorem 4.1.3.

For s = r + 3.

$$E((X(r+3,n,m,k) - X(r,m,n,k))|X(,n,m,k) = x)$$

$$= \frac{\gamma_{r+1}\gamma_{r+2}\lambda_{r+3}}{\lambda^3} \frac{1}{(m+1)^2} \sum_{j=0}^{2} \frac{2}{\Gamma(j+1)\Gamma(3-j)} (-1)^j \left(\frac{1}{\gamma_{r+3-j}}\right)^2$$

$$= \frac{1}{\lambda^3} \frac{1}{(m+1)^2} \left[\frac{\gamma_{r+1}\gamma_{r+2}}{2\lambda_{r+3}} - \frac{\gamma_{r+1}\lambda_{r+3}}{\gamma_{r+2}} + \frac{\gamma_{r+2}\lambda_{r+3}}{2\gamma_{r+1}}\right]$$

$$= \frac{1}{\lambda^3} \frac{1}{(m+1)^2} [\frac{1}{2}\lambda_{r+3} + \frac{3}{2}(m+1) + \frac{(m+1)^2}{\lambda_{r+3}}\lambda_{r+2} + \frac{(m+1)^2}{\lambda_{r+2}} +$$

$$\frac{1}{2}\lambda_{r+1}$$

$$+\frac{1}{2}(m+1) + \frac{(m+1)^2}{\lambda_{r+1}}]$$

$$= \frac{1}{\lambda_{r+3}} + \frac{1}{\lambda_{r+2}} + \frac{1}{\lambda_{r+1}}.$$

It can be shown that

$$E((X(r+s,n,m,k) - X(r,m,n,k))|X(,n,m,k) = x)$$

$$= \frac{1}{\lambda_{r+s}} + \frac{1}{\lambda_{r+s-1}} + \cdots + \frac{1}{\lambda_{r+1}}.$$

Thus from Theorem 4.1.3, we obtain the following Theorem.

Theorem 4.1.4.

Let the random variable X have an absolutely continuous cdf F(x) = 0 and pdf f(x). We assume E(X) exist. Then for r, $1 < r < s \leq n$.

$$E(X(s,n,m,k) - X(r,n,m,k)^p|X(r,n,m,k) = x)$$

$$= \frac{1}{\lambda_{r+s}} + \frac{1}{\lambda_{r+s-1}} + \cdots + \frac{1}{\lambda_{r+1}}, \quad m \neq -1, \tag{4.1.3}$$

If and only if $F(x) = 1 - e^{-\lambda x}$, $\lambda > 0$ and $x \geq 0$.

Theorem 4.1.5.

Let the random variable X have an absolutely continuous cdf F(x) and pdf f(x). We assume $E(X^m)$, for a positive integer m, exist. Then for $1 < r < s \leq n$,

$$E(X(s,n,m,k)^m | X(r,n,m,k) = x)$$
$$= x^m + \frac{mx^{m-1}}{\gamma_{r+1}} + \frac{m(m-1)x^{m-2}}{\gamma_{r+1}^2} + \frac{m(m-1)(m-2)x^{m-3}}{\gamma_{r+1}^3} + \dots + \frac{m!}{\gamma_{r+1}^{m-1}} X +$$
$$\frac{m!}{\gamma_{r+1}^m}$$

If and only if $F(x) = 1 - e^{-x}$, and $x \geq 0$.

Proof

$$E(X(s,n,m,k)^m | X(r,n,m,k) = x) = \gamma_{r+1} \int_x^\infty y^m \frac{(1-F(y))^{\gamma_{r+1}-1}}{(1-F(x))^{\gamma_{r+1}}} f(y)$$

Suppose that $F(x) = 1 - e^{-x}$, and $x \geq 0$, then

$$E(X(s,n,m,k)^m | X(r,n,m,k) = x) = \gamma_{r+1} \int_x^\infty y^m \frac{e^{-\gamma_{r+1}y}}{e^{\gamma_{r+1}x}} dy$$
$$= x^m + \frac{mx^{m-1}}{\gamma_{r+1}} + \frac{m(m-1)x^{m-2}}{\gamma_{r+1}^2} + \frac{m(m-1)(m-2)x^{m-3}}{\gamma_{r+1}^3} + \dots + \frac{m!}{\gamma_{r+1}^{m-1}} X +$$
$$\frac{m!}{\gamma_{r+1}^m},$$

Suppose that

$$E(X(s,n,m,k)^m | X(r,n,m,k) = x)$$

$$= x^m + \frac{mx^{m-1}}{\gamma_{r+1}} + \frac{m(m-1)x^{m-2}}{\gamma_{r+1}^2} + \frac{m(m-1)(m-2)x^{m-3}}{\gamma_{r+1}^3} + \dots + \frac{m!}{\gamma_{r+1}^{m-1}} X + \frac{m!}{\gamma_{r+1}^m},$$

then,

$$\gamma_{r+1} \int_x^\infty y^m \frac{(1-F(y))^{\gamma_{r+1}-1}}{(1-F(x))^{\gamma_{r+1}}} f(y) dy$$

$$= x^m + \frac{mx^{m-1}}{\gamma_{r+1}} + \frac{m(m-1)x^{m-2}}{\gamma_{r+1}^2} + \frac{m(m-1)(m-2)x^{m-3}}{\gamma_{r+1}^3} + \dots + \frac{m!}{\gamma_{r+1}^{m-1}} X + \frac{m!}{\gamma_{r+1}^m},$$

i.e.:

$$\gamma_{r+1} \int_x^\infty y^m (1 - F(y))^{\gamma_{r+1}-1} f(y) dy$$

$$= (1 - F(x))^{\gamma_{r+1}} (x^m + \frac{mx^{m-1}}{\gamma_{r+1}} + \frac{m(m-1)x^{m-2}}{\gamma_{r+1}^2} + \frac{m(m-1)(m-2)x^{m-3}}{\gamma_{r+1}^3} +$$

$$\dots + \frac{m!}{\gamma_{r+1}^{m-1}} X + \frac{m!}{\gamma_{r+1}^m})$$

Differentiating both sides of the above equation with respect to x, we obtain on simplification

$$\frac{f(x)}{1-F(x)} = 1.$$

Using the boundary conditions $F(0) = 0$ and $F(\infty) = 1$, we obtain

$F(x) = 1 - e^{-x}$, and $x \geq 0$.

4.2. CHARACTERIZATIONS
BY INDEPENDENCE PROPERTIES

Theorem 4.2.1.

Let X be a non-negative random variable having an absolutely continuous (with respect to Lebesgue measure) strictly increasing cdf F(x) with F(0) = 0 and F(x) < 1 for all x > 0. Then the following two properties are equivalent

   a) X has an exponential distribution, $E(0, \sigma^{-1})$
   b) for $1 < r \leq n$, the statistics {X(r,n,m,k) - X(r-1,n,m,k)} and X(r-1, n,m,k) are independent.

Proof

Integrating out $x_1, \ldots, x_{r-2}, x_{r+1}, \ldots, x_n$, and using the transformation

$$U = X(r\text{-}1,n,m,k) \text{ and } W = (X(r,n,m,k) - X(r\text{-}1,n,m,k)),$$

we get on simplification the joint pdf $f_{UW}(u,w)$ of U and W as

$$f_{UW}(u,w) =$$

$$\frac{c_{r-2}}{(r-2)!}(1 - F(u))^m g_m^{r-2}(F(u))(1 - F(u+w))^{\gamma_r - 1} f(u) f(u+w),$$

$$(4.2.1)$$

for $0 < u, w < \infty$, and $f_{UW}(u,w) = 0$, otherwise, where

$$c_{r-1} = \prod_{j=1}^{r} \gamma_j, \, r = 1,2,\ldots,n, \, \gamma_j = k + (n\text{-}j)(m + 1), \, j = 1,2,\ldots,n;$$

$$g_m(x) = \frac{1}{m+1}(1 - (1 - x)^{m+1}), \text{ if } m \neq -1,$$

and

$g_m(x) = -\ln(1-x)$, if $m = -1$.

Let X have the exponential $E(0, \sigma^{-1})$ distribution. Then on simplification, we get from (4.1.1.) that

$$f_{UW}(am) = \frac{c_{r-2}}{\sigma^2(r-2)!}\, g_m^{r-2}(1 - e^{-u/\sigma})e^{-\gamma_{r-1}u/\sigma}e^{-\gamma_r w/\sigma}, \quad u > 0, w > 0.$$

$$(4.2.2)$$

It follows from (4.2.2) that in this case W and U are independently distributed.

We will prove here of $(b) \Rightarrow (a)$.

The pdf $f_U(u)$ of the random variable $U = X(r-1, n, m, k)$ is

$$f_U(u) = \frac{c_{-2}}{(r-2)!}\,(1 - F(u))^{-1+\gamma_{-1}}\, g_m^{r-2}(F(u))\, f(u) \qquad (4.2.3)$$

Using the relation $\gamma_{r-1} = \gamma_r + m + 1$, we get the conditional distribution of $f_{W|U}(W|\, U = u)$ as

$$f_{W|U}(w|\, U = u) = \left((\bar{F}(u + w))/(\bar{F}(u))\right)^{\gamma_r - 1} f(u + w)(\bar{F}(u))^{-1}$$

$$(4.2.4)$$

for $w > 0, u > 0$, where $\bar{F}(u) = 1 - F(u)$.

Integrating the expression in (4.2.4) with respect to w, we obtain that the conditional probability $P\{W > w|U = u\}$ has the form

$$\bar{F}_{W|w}(w_1|U = u) = (\bar{F}(u + w)/(\bar{F}(u)))^{\gamma_r}/\gamma_r, \quad u > 0, w > 0.$$

$$(4.2.5)$$

Suppose that W and U are independent. Then we get from (4.2.5) that

$$(\bar{F}(u + w)/(\bar{F}(u)))^{\gamma_r} = G(w), \text{ for all } u > 0 \text{ and } w > 0,$$

where $G(w)$ is a function of w only. Now taking limit $u \to 0$, we get that

$$G(w) = \left(\bar{F}(w)\right)^{\gamma_r}.$$

Hence for all $w > 0$ and $u > 0$, we obtain

$$\bar{F}(u+w) = \bar{F}(u)\,\bar{F}(w).\qquad\qquad(4.2.6)$$

The solution of equation (4.2.6) is (see Aczel [1966])

$$\bar{F}(u) = e^{-u\sigma^{-1}},$$

where σ is an arbitrary positive number. Since $F(x) = 1 - \bar{F}(x)$ is a distribution function, σ must be positive.

If $k = 1$ and $m = 0$, then from Theorem 42.1, we obtain the result of Rossberg (1972) for order statistics. If $k = 1$ and $m = -1$, then Theorem 4.2.1 presents the result of Ahsanullah (1978) giving the corresponding characterizations of the exponential distribution by the independence of record statistics $X(r-1)$ and $X(r)-X(r-1)$.

Remark 4.2.1.

From Theorem 4.2.1 it is easy to see that for the exponential $E(0,\sigma)$ distribution, the regression function $\varphi(x) = E(X(r + 1,m,n,k)|X(r,n,n,k) = x)$ is a linear function of x. It can be shown that this is a characteristic property of the exponential distribution.

4.3. CHARACTERIZATIONS BY IDENTICAL DISTRIBUTIONS

To prove the next theorem we need the monotonicity of a failure rate (hazard rate). We define the failure rate of a random variable X as $h(x) = f(x)/(1-F(x))$, where $F(x)$ is the cdf and $f(x)$ is the pdf of X. We define the normalized spacing of gos as

$$D(r + 1, \ n,m,k) = \gamma_{r+1}(X(r + 1, n, m, k) - X(r, n, m, k), r = 0,1,2 \ldots, n - 1$$

Theorem 4.3.1.

Let X be a non-negative r.v. having an absolutely continuous (with respect to Lebesgue measure) strictly increasing distribution function $F(x)$ for all $x > 0$, $F(0) = 0$ and $F(x) < 1$ for all $x > 0$. Then the following properties are equivalent.

(a) The random variable X has an exponential distribution with pdf
$f(x) = \sigma^{-1}\exp(-x/\sigma)$, $0 < x < \infty$, $\sigma > 0$;
(b) X has increasing or decreasing failure rate and there exist integers
r and n, $1 \le r < n$, such that the normalized spacing $D(r,n,m,k)$ and
$D(r + 1,n,m,k)$, are identically distributed for some $m \ge -1$.

Proof
It is easy to show that (a) $\Rightarrow$ (b). We will prove that (b) $\Rightarrow$ (a).
It follows from (4.3.9) that the pdf of $D(r) = D(r,n,m,k)$ for $r \ge 2$ can be written as

$$f_{D(r)}(x) = \frac{c_{r-2}}{(r-2)!} \int_0^\infty (\overline{F}(y))^m f(y) g_m^{r-2}(F(y)) (\overline{F}(y + \frac{x}{\gamma_r}))^{\gamma_r - 1} f(y + \frac{x}{\gamma_r}) dy$$

$$(4.3.1)$$

Then the cdf of D(r,n,m,k) is

$$F_{D(r)}(x) = 1 - \frac{c_{r-2}}{(r-2)!} \int_x^\infty \int_0^\infty (\bar{F}(y))^m f(y) g_m^{r-2}(F(y))(\bar{F}(y +$$

$$\frac{u}{\gamma_r}))^{\gamma_r-1} f(y + \frac{u}{\gamma_r}) dy du =$$

$$1 - \frac{c_{r-2}}{(r-2)!} \int_0^\infty (\bar{F}(y))^m f(y) g_m^{r-2}(F(y)) \int_x^\infty (\bar{F}(y +$$

$$\frac{u}{\gamma_r}))^{\gamma_r-1} f(y + \frac{u}{\gamma_r}) du \, dy =$$

$$1 - \frac{c_{r-2}}{(r-2)!} \int_0^\infty (\bar{F}(y))^m f(y) g_m^{r-2}(F(y))(\bar{F}(y + \frac{x}{\gamma_r}))^{\gamma_r} dy \qquad (4.3.2)$$

and analogously the cdf of D(r + 1,n,m,k) can be written respectively as

$$F_{D(r+1)}(x) = 1 - \frac{c_{r-1}}{(r-1)!} \int_0^\infty (\bar{F}(y))^m f(y) g_m^{r-1}(F(y))(\bar{F}(y +$$

$$\frac{x}{\gamma_{r+1}}))^{\gamma_{r+1}} dy. \qquad (4.3.3)$$

Taking into account the definition of the function g_m one can see that the RHS of (4.3.3) can be transformed as follows:

$$1 - \frac{c_{r-2}}{(r-2)!} \int_0^\infty (\bar{F}(y))^m f(y) g_m^{r-2}(F(y))(\bar{F}(y + \frac{x}{\gamma_r}))^{\gamma_r} dy =$$

$$1 - \frac{c_{r-2}}{(r-1)!} \int_0^\infty (\bar{F}(y + \frac{x}{\gamma_r}))^{\gamma_r} d(g_m^{r-1}(F(y))) =$$

$$1 - \frac{\gamma_r c_{r-2}}{(r-1)!} \int_0^\infty g_m^{r-1}(F(y))(\bar{F}(y + \frac{x}{\gamma_r}))^{\gamma_r-1} f(y + \frac{x}{\gamma_r}) dy =$$

$$1 - \frac{c_{r-1}}{(r-1)!} \int_0^\infty g_m^{r-1}(F(y))(\bar{F}(y + \frac{x}{\gamma_r}))^{\gamma_r-1} f(y + \frac{x}{\gamma_r}) dy. \qquad (4.3.4)$$

Let statement (b) of the theorem be valid. Moreover suppose that F has increasing failure rate h(x). Then $\ln \bar{F}$ is concave and for $m \geq -1$, we have

$$\ln(\bar{F}(y + \frac{x}{\gamma_r})) = \ln(\bar{F}(\frac{(m+1)y}{\gamma_r} + \frac{\gamma_{r+1}}{\gamma_r}(y + \frac{x}{\gamma_{r+1}}))$$

$$\geq \frac{m+1}{\gamma_r}\ln(\bar{F}(y)) + \frac{\gamma_{r+1}}{\gamma_r}\ln\bar{F}(y + \frac{x}{\gamma_{r+1}}).$$

Thus,

$$(\bar{F}(y + \frac{x}{\gamma_r}))^{\gamma_r} \geq (\bar{F}(y))^{m+1}(\bar{F}(y + \frac{x}{\gamma_{r+1}}))^{\gamma_{r+1}}. \tag{4.3.5}$$

Since X(r,n,m,k) and X(r + 1,n,m,k) are identically distributed and $\gamma_r = \gamma_{r+1} + m + 1$, we get from (4.3.5), that

$$0 \quad = \quad \frac{c_{r-1}}{(r-1)!}\int_0^\infty g_m^{r-1}(F(y))[(\bar{F}(y))^m f(y)(\bar{F}(y + \frac{x}{\gamma_{r+1}}))^{\gamma_{r+1}-}$$

$$(\bar{F}(y + \frac{x}{\gamma_r}))^{\gamma_r - 1} f(y + \frac{x}{\gamma_r})]dy \leq$$

$$\leq \frac{c_{r-1}}{(r-1)!}\int_0^\infty g_m^{r-1}(F(y))[(\bar{F}(y))^{m+1}(\bar{F}(y + \frac{x}{\gamma_{r+1}}))^{\gamma_{r+1}}[h(y) -$$

$$h(y + \frac{x}{\gamma_{r+1}})] \, dy \tag{4.3.6}$$

Since

$$h(y) \geq h(y + x/y_{r+1})$$

for any x > 0, the RHS of (4.3.6) takes zero value provided that

$$h(y) - h(y + \frac{x}{\gamma_{r+1}}) = 0$$

for almost all y > 0 and all x > 0. The constancy of the failure rate characterizes the exponential distribution. Since additionally F(0) = 0 and

F is a strictly increasing for x > 0 distribution function, we obtain that $F(x) = 1-e^{-x/\sigma}$, $x \geq 0$, where σ is any arbitrary positive number. The case, when h(x) is a decreasing failure rate can be considered analogously.

Remark 4.3.1.

The condition of monotone failure rate in the Theorem 4.3.1 can be replaced by NBU and NWU properties. A distribution function F is said to be NBU (NWU) if $\bar{F}(x + y) \leq (\geq)\bar{F}(x)\bar{F}(y)$ for all x,y and x + y within the support of F.

Remark 4.3.2.

The condition of identical distribution of the normalized spacings of generalized order statistics in Theorem 4.3.2 can be replaced by the equality of their expectations.

The proof of the following two theorems are similar to that of Theorem 4.2.1.

Theorem 4.3.2.

Let X be a non negative random variable having an absolutely continuous (with respect to Lebesgue measure) strictly increasing distribution function F(x) for all x > 0, F(0) = 0 and F(x) < 1 for all x > 0. Then the following two properties are equivalent.

a) X has an exponential distribution with the pdf $f(x) = \sigma^{-1}\exp(-x/\sigma)$,x > 0, $\sigma > 0$;

b) X has a monotone failure rate and for one k, one r , one n and $m \geq -1$, the statistics $D(r,n,m,k)$ and $\gamma_r X(1,n-r + 1,m,k)$ are identically distributed.

Theorem 4.3.3.

Let X be a non-negative random variable having an absolutely continuous (with respect to Lebesgue measure) strictly increasing distribution function F(x) for all x > 0 and F(x) < 1 for all x > 0. Then the following properties are equivalent.

a) X has an exponential distribution with the pdf $f(x) = \sigma^{-1}\exp(-x/\sigma)$, $0 < x < \infty$, $\sigma > 0$;

b) X has a monotone failure rate and for one k, one r, one n and m$\geq$ -1, (k $\geq$ 1, m is a real number, n and rare integers , 1 < r$\leq$n) the normalized spacing {k + (n-r)(m + 1)}{X(r,n,m,k) - X(r-1,n,m,k)}

and random variable X are identically distributed.

APPENDIX

Assumptions A

Suppose the random variable X is absolutely continuous with the cumulative distribution function $F(x)$ and the probability density function $f(x)$. We assume that $\alpha=\inf\{x|F(x)>0\}$ and $\beta = sup\{x\ |F(x)<1\}$. We also assume that $f(x)$ is a differentiable for all x, $-\alpha<x<\beta$ and $E(X)$ exists.

Lemma 2.1.

Under the Assumption A, if $E(X|X\le x)=g(x)\tau(x)$, where $\tau(x)=\dfrac{f(x)}{F(x)}$ and $g(x)$ is a continuous differentiable function of x with

the condition that $\dfrac{\int_\alpha^x uf(u)du}{F(x)}$ is finite for x>α, then f(x) $=ce^{\int \frac{x-g'(x)}{g(x)}dx}$,

where c is a constant determined by the condition $\int_\alpha^\beta f(x)dx = 1$.

Lemma 2.2.

Under the Assumption 2.1, if $E(X\}X \geq x) = h(x)r(x)$, where $r(x) = \dfrac{f(x)}{1 - F(x)}$ and h(x) is a continuous differentiable function of x ,- $\alpha < x < \beta$, with the condition that $\int_x^\beta \dfrac{u + h\prime(u)}{h(u)} du$ is finite for x, $\alpha < x < \beta$, then f(x) $= ce^{-\int \frac{x + h\prime(x)}{h(x)} dx}$, where c is a constant determined by the condition $\int_\alpha^\beta f(x) dx = 1$.

For Proof, see Ahsanullah (2016) pages 49-50.

REFERENCES

Abdel-Aty, S. H. (1954). Ordered variables in discontinuous distributions. *Statististica Neerlandica* 8, 61-82.

Abo-Eleneen, Z. A. (2001). *Information in order statistics and their concomitants and applications*. Ph.D. dissertation, Zagazig Univ., Egypt.

Abramowitz, M. and Stegan, I. (1972). *Handbook of Mathematical Functions*. Dover, New York, NY.

Aczel, J. (1966). *Lectures on Functional Equations and Their Applications*. Academic Press, New York, NY.

Adatia, A. and Chan, L. K. (1981). Relations between stratified, grouped and selected order statistics samples. *Scand. Actuar. J.*, 4, 193-202.

Adke, S. R. (1993). Records generated by Markov sequences. *Statist. and Prob. Letters,* 18, 257-263.

Afify, A. (2013). *Characterizations of Probability Distributions*, Lambert Academic Publishers, Deutschland, Germany.

Aggarwal, M. L. and Nagabhushanam, A. (1971). Coverage of a record value and related distribution problems. *Bull. Calcutta Stat. Assoc.*, 20, 99-103.

Ahsanullah M. and Nevzorov V. (1996a). Distributions of order statistics generated by records. *Zapiski Nauchn. Semin. POMI,* 228, 24-30 (in Russian). English transl. in J. Math. Sci.

Ahsanullah M. and Nevzorov V. (1997). One limit relation between order statistics and records. *Zapiski nauchnyh seminarov POMI* (Notes of Sci Seminars of POMI), v. 244, 218-226 (in Russian).

Ahsanullah M. and Nevzorov V. (2000). Some distributions of induced records. *Biometrical Journal,* 42, 153-165.

Ahsanullah M. and Nevzorov V. (2001c). Extremes and records for concomitants of order statistics and record values. *J. of Appl. Statist. Science*, v. 10,181-190.

Ahsanullah M. and Nevzorov V. (2004). Characterizations of distributions by regressional properties of records. *J. Appl. Statist. Science*, v. 13, N1, 33-39.

Ahsanullah M. and Nevzorov V. (2011). Record statistics. *International Encyclopedia of Statistical Science*, part 18, 1195-1202.

Ahsanullah, M. (1975). A characterization of the exponential distribution. In: G. P. Patil, S. Kotz and J. Ord. eds., *Statistical Distributions in Scientific Work*, Vol. 3, 71-88., D. Reidel Publishing Company, Dordrecht-Holland..

Ahsanullah, M. (1976). On a characterization of the exponential distribution by order statistics. *J. Appl. Prob.*,13, 818-822.

Ahsanullah, M. (1977). A characteristic property of the exponential distribution. *Ann. of Statist.,* 5, 580-582.

Ahsanullah, M. (1978a). A characterization of the exponential distribution by spacing. *J. Prob. Appl.*, 15, 650-653.

Ahsanullah, M. (1978b). On characterizations of exponential distribution by spacings. *Ann. Inst. Stat. Math.*, 30, A, 163-166.

Ahsanullah, M. (1978c). Record values and the exponential distribution. *Ann. Inst. Statist. Math.*, 30, A, 429-433.

Ahsanullah, M. (1979). Characterization of the exponential distribution by record values. *Sankhya*, 41, B, 116-121.

Ahsanullah, M. (1980). Linear prediction of record values for the two parameter exponential distribution. *Ann. Inst. Stat. Math.*, 32, A, 363-368.

Ahsanullah, M. (1981a). Record values of the exponentially distributed random variables. *Statistiche Hefte*, 2, 121-127.

Ahsanullah, M. (1981b). On a Characterization of the Exponential Distribution by Weak Homoscedasticity of Record Values. *Biom. J.*, 23, 715-717.

Ahsanullah, M. (1981c). On characterizations of the exponential distribution by spacings. *Statische Hefte*, 22, 316-320.

Ahsanullah, M. (1982). Characterizations of the Exponential Distribution by Some Properties of Record Values. *Statistiche Hefte*, 23, 326-332.

Ahsanullah, M. (1984). A characterization of the exponential distribution by higher order gap. *Metrika,* 31, 323-326.

Ahsanullah, M. (1986a). Record values from a rectangular distribution. *Pakistan J. Statist.*, A2, 1-5.

Ahsanullah, M. (1986b). Estimation of the parameters of a rectangular distribution by record values. *Comp. Stat. Quarterly,* 2, 119-125.

Ahsanullah, M. (1987a). Two Characterizations of the Exponential Distribution. *Comm. Statist. Theory-Methods*, 16, no. 2, 375-381.

Ahsanullah, M. (1987b). Record Statistics and the Exponential Distribution. *Pak. J. Statist.*, 3, A, 17-40.

Ahsanullah, M. (1988a). On a conjecture of Kakosian, Klebanov and Melamed. *Statistiche Hefte,* 29, 151-157.

Ahsanullah, M. (1988b). *Introduction to Record Statistics,* Ginn Press, Needham Heights, MA.

Ahsanullah, M. (1988c). Characteristic properties of order statistics based on a random sample size from an exponential distribution. *Statistica Neerlandica*, 42, 193-197.

Ahsanullah, M. (1990). Estimation of the parameters of the Gumbel distribution based on m record values. *Statistische Hefte*, 25, 319-327.

Ahsanullah, M. (1991). Some Characteristic Properties of the Record Values from the Exponential Distribution. *Sankhya, B* 53, 403-408.

Ahsanullah, M. (1992). Record Values of Independent and Identically Distributed Continuous Random Variables. *Pak. J. Statist.*, 8, no. 2, A, 9-34.

Ahsanullah, M. (1994). Records of Univariate Distributions. *Pak. J. Statist.*, 9, no. 3, 49-72.

Ahsanullah, M. (1995). *Record Statistics*. Nova Science Publishers Inc., New York, NY.

Ahsanullah, M. (2000). Generalized order statistics from exponential distribution. *J. Statist. Plann. and Inf.*, 25, 85-91.

Ahsanullah, M. (2003). Some characteristic properties of the generalized order statistics from exponential distribution. *J. Statist. Research*, vol. 37, 2, 159-166.

Ahsanullah, M. (2004). *Record Values-Theory and Applications.* University Press of America, Lanham, MD.

Ahsanullah, M. (2006). The generalized order statistics from exponential distribution. *Pak. J. Statist.,* vol. 22, 2, 121-128.

Ahsanullah, M. (2007).Some characterizations of the power function distribution based on lower generalized order statistic. *Pak. J. Stat.,* vol. 23, 2007, 139-146.

Ahsanullah, M. (2008).Some characteristic properties of weak records of geometric distributions. *Journal of Statistical Theory and Applications,* vol 7 no.1, 81-92.

Ahsanullah, M. (2009 b) Records and Concomitants. *Bulletin of the Malaysian Mathematical Sciences Society,* vol. 32, no. 2,101-117.

Ahsanullah, M. (2009a). On characterizations of geometric distribution by weak records. *Journal of Statistical Theory and Applications*, vol. 8, no. 1, 5-12.

Ahsanullah, M. (2009c) On some characterizations of univariate distributions based on truncated moments of order statistics. *Pakistan Journal of Statistics*, vol. 25, no.2, 83-91. Corrections. *Pakistan Journal of Statistics*, vol. 28, no.3, 2010, 563.

Ahsanullah, M. (2010). Some Characterizations of Exponential Distribution by Upper Record Values. *Pakistan Journal of Statistics,* vol. 26, no.1, 69-75.

Ahsanullah, M. (2013a). Inferences of Type II extreme value distribution based on record values. *Applied Mathematical Sciences*, 7(72), 3569-3578.

Ahsanullah, M. (2013b). On generalized Type I logistic distribution. *Afrika Statistics, 8*, 579-585.

Ahsanullah, M. (2016). *Characterizations of Univariate Distributions*. Atlantis-Press, Paris, France.

Ahsanullah, M. and Aliev, F. (2011). A characterization of geometric distribution based on weak records. *Stat. Papers*. 52,651-655.

Ahsanullah, M. and Aliev, F. (2008). Some characterizations of exponential distribution by record values. *Journal of Statistical Research*, Vol. 2, No. 1, 11-16.

Ahsanullah, M. and Anis, M. Z. (2017). Some characterizations of exponential distribution. *International Journal of Statistics and Probability,* vol. 6. no. 5,1-8.

Ahsanullah, M. and Hamedani, G. G. (2013). Characterizations of continuous distributions based on conditional expectations of generalized order statistics. *Communications in Statistics, Theory, and Methods*, 42, 3608-3613.

Ahsanullah, M. and Hijab, O (2006). Some characterizations of geometric distribution by weak record. In *Recent Developments in Ordered Random Variables*. Nova Science Publishers Inc., 187-195. Edited by M. Ahsanullah and M. Z. Raqab.

Ahsanullah, M. and Holland, B. (1984). Record values and the geometric distribution. *Statistische Hefte*, 25, 319-327.

Ahsanullah, M. and Holland, B. (1987). Distributional properties of record values from the geometric distribution. *Statistica Neerlandica*, 41, 12-137.

Ahsanullah, M. and Kirmani, S. N. U. A. (1991). Characterizations of the Exponential Distribution by Lower Record Values. *Comm. Statist. Theory-Methods*, 20(4), 1293-1299.

Ahsanullah, M. and Nevzorov, V. B. (2001a). *Ordered Random Variables*. New York, NY: Nova Science Publishers Inc.

Ahsanullah, M. and Nevzorov, V. B. (2001b). Distribution between uniform and exponential. In: M. Ahsanullah, J. Kenyon and S. K. Sarkar eds., *Applied Statistical Science* IV, 9-20.

Ahsanullah, M. and Nevzorov, V. B. (2005). *Order Statistics. Examples and Exercise*. New York, NY: Nova Science Publishers.

Ahsanullah, M. and Raqab, M. Z. (2006) *Bounds and Characterizations of Record Statistics*. Nova Science Publishers Inc. New York, NY, USA.

Ahsanullah, M. and Raqab, M. Z. (2007). *Recent Developments in Ordered Random Variables*. Nova Science Publishers Inc. New York, NY, USA.

Ahsanullah, M. and Shakil, M, (2011a). . On Record Values of Rayleigh distribution. *International Journal of Statistical Sciences*. Vol. 11 (special issue), 2011, 111-123.

Ahsanullah, M. and Shakil, M. (2011 b). Record values of the ratio two exponential distribution. *Journal of Statistical Theory and Applications*. vol. 10, no,3, 393-406.

Ahsanullah, M. and Shakil, M. (2012). A note on the characterizations of Pareto distribution by upper record values. Commun. *Korean Mathematical Society*, 27 (4), 835-842.

Ahsanullah, M. and Tsokos, C. P. (2005). Some distributional properties of Record Values from Generalized Extreme Value Distributions. *Journal of Statistical Studies*. Vol. 25, 11-18.

Ahsanullah, M. and Wesolowski, J. (1998). *Linearity of Best Non-Adjacent Record Values*. Sankhya, B, 231-237.

Ahsanullah, M. and Yanev, G. P. (2008). *Records and Branching Processes*. Nova Science Publishers Inc. New York, NY, USA.

Ahsanullah, M., Aliev, F., and Oncel, S. Y. (2013). A note on the characterization of Pareto distribution by the hazard rate of upper record values. *Pakistan Journal of Statistics*, 29(4), 447-452.

Ahsanullah, M., Alzaid, A. A. and Golam Kibria, B. M. (2014). On the residual life of the k out of n system. *Bulletin of the Malaysian Mathematical Sciences Society*, 2 (37-1), 83-91.

Ahsanullah, M., Hamedani, G. G. and Shakil, M. (2010). Expanded Version of Record Values of Univariate Exponential Distribution. *Technical Report Number 479*, Marquette University.

Ahsanullah, M., Hamedani, G. G. and Shakil, M. (2010a) On Record Values of Univariate Exponential Distributions. *Journal of Statistical Research* vol, 44, No. 2, 267-288.

Ahsanullah, M., Hamedani, G. G. and Shakil, M. (2010b). On Record Values of Univariate Exponential Distributions. *Journal of Statistical Research* vol. 44, No. 2, 267-288.

Ahsanullah, M., Nevzorov, V. B. and Yanev, G. P. (2010) Characterizations of distributions via order statistics with random exponential shift. *Journal of Applied Statistical Science*, 18,3,. 297-305.

Ahsanullah, M., Nevzorov, V. P., and Shakil, M. (2013). *An Introduction to Order Statistics*. Paris, France: Atlantis Press.

Ahsanullah, M., Shah, I. A. and Yanev, G. P. (2013). On characterizations of exponential distribution through order statistics and record values with random shifts. *J Stat Appl Pr,* 2(3), 223-227.

Ahsanullah, M., Shakil, J., and Golam Kibria, B. M. (2014). A Note on a characterization of Gompertz-Verhulst distribution. *J. Stat. Theory and Applications*, 13(1), 17-26.

Ahsanullah, M., Shakil, M., and Golam Kibria, B. M. (2013). A characterization of power function distribution based on lower records. *ProbStat Forum*, 6, 68-72.

Ahsanullah, M., Shakil, M., and Golam Kibria, B. M. (2013). A Note on a probability distribution with fractional moments arising from generalized Person system of differential equation. *International Journal of Advanced Statistics and Probability,* 1 (3), 132-141.

Ahsanullah, M., Yanev, G. P., and Onica, C. (2012). Characterizations of logistic distribution through order statistics with independent exponential shifts. *Economic Quality Control*, 27(1), 85-96.

Akbaria, M. and Fashandia, M. (2014). On characterization results based on the 7 number of observations near the k-records. *Statistics,* 48, no. 3, 633-640.

Akhundov, I. and Nevzorov, V. (2006). Conditional distributions of record values and characterizations of distributions. *Probability and Statistics. 10 (Notes of Sci. Semin. POMI, v. 339*), 5-14.

Akhundov, I. and Nevzorov, V. (2008). Characterizations of distributions via bivariate regression on differences of records. In: *Records and Branching Processes.* Nova Science Publishers, 27-35.

Akhundov, I., Berred, A. and Nevzorov V. (2007). On the influence of record terms in the addition of independent random variables. *Communications in Statistics: Theory and Methods*, v. 36, n. 7, 1291-1303.

Aliev, A. A. (1998), Characterizations of discrete distributions through week records. *Journal of Applied Statistical Science,* Vol 8(1) 13-16.

Aliev, A. A. (1998). New characterizations of discrete distributions through weak records. *Theory of Probability and Applications*, 44, no. 4, 415-421.

Aliev, A. A. and Ahsanullah, M. (2002). On characterizations of discrete distrbutions by regression of record values. *Pakistan Journal of Statistics*, Vol, 18 (3), 315-421.

Alpuim, M. T. (1985). Record values in populations with increasing or random dimension. *Metron*, 43, no. 3-4, 145-155.

Andel, J. (1990). Records in an AR (1) process. *Ricerche Mat.,* 39, 327-332.

Arnold, B. C. (1971). *Two characterizations of the exponential distribution*. Technical Report, Iowa State University.

Arnold, B. C. (1980). Two characterizations of the geometric distribution. *J. Appl. Prob.* 17, 570-573.

Arnold, B. C. (1983). *Pareto distributions.* International Cooperative Publishing House, Fairland, Maryland.

Arnold, B. C. (1985). *p*-Norm bounds of the expectation of the maximum of possibly dependent sample. *J. Multiv. Anal.* 17, 316-32.

Arnold, B. C. (1988). Bounds on the expected maximum. *Commun. Statist-Theory Meth.* 17, 2135-50.

Arnold, B. C. and Balakrishnan, N. (1989). *Relations, Bounds and Approximations for Order Statistics.* Lecture Notes in Statistics, No. 53, Springer-Verlag, New York, NY.

Arnold, B. C. and Ghosh, M. (1976). A characterization of geometric distributions by distributional properties of order statistics. *Scan. Actuarial J.*, 232-234.

Arnold, B. C. and Groeneveld, R. A. (1979). Bounds on expectations of linear systematic statistics based on dependent samples. *Ann. Statist.* 7, 220-3. Correction 8, 1401.

Arnold, B. C. and Nagaraja, H. N. (1991). Lorenz ordering of exponential order statistics. *Statist. Probab. Lett.* 11, 485-90.

Arnold, B. C. and Villasenor, J. A. (1998). The asymptotic distributions of sums of records. *Extremes*, 1, no.3, 351-363.

Arnold, B. C. and Villasenor, J. A. (2013). Exponential characteriztions motivated by the structure of order statistics in samples of size 2. *Statist. Probab. Lett.* 83,596-601., 485-90.

Arnold, B. C., Balakrishnan, N. and Nagaraja, H. N. (1998). *Records.* John Wiley& Sons Inc., New York. NY.

Arnold, B. C., Castillo, E. and Sarabia, J. M. (1999). *Conditional specification of statistical models.* Springer, New York, NY.

Athar, H. and Akhter, Z. (2015). Some characterization of continuous distributions based on order statistics. *International Journal of Computational and Theoretical Statistics*, 291), 31-36.

Athar, H. and Akhter, Z. (2015). Some characterization results based on conditional expectation of dual generalized order. *Statistics. Probstats Forum*, 8, 103-111.

Athar, H. and Zubadah-e-Noor (2016). Characterizing probability distributions by conditional expectation of order statistics, *Mathmatical Science Letter*, 5(3),323-338.e.

Athar, H. Haque, Z. and Khan, R. U. (2010). On characterization of probability distributions through generalized order statistics, *Journal of Indian Society for Probability and Statistics*. 12,102-112.

Athar, H., Yaqub, M. and Islam, H. M. (2003). *Aligarh Journal of Statistics*, 23, 97-105.

Athreya, J. S. and Sethuraman, S. (2001). On the asymptotics of discrete order statistics. *Statist. Probab. Lett*. 54, 243-9.

Aven, T. (1985). Upper (lower) bounds as the mean of the maximum (minimum) of several random variables. *J. Appl. Probab*. 22, 723-8.

Azlarov, T. A. and Volodin, N. A. (1986). *Characterization problems associated with the exponential distribution*. Springer-Verlag, New York, NY

Bairamov, I. and Ozkal, T. (2007). On characterizations of distributions through the properties of conditional expectations of order statistics. *Commun. In Statist. Theory- Methods,* 36, 1319-1326.

Bairamov, I. and Stepanov, A. (2011). Numbers of near bivariate record-Concomitant observations. *Journal* of *Multivariate Analysis,* 102, 908-917.

Bairamov, I. and Stepanov, A. (2013). Numbers of near-maxima for F^{α} -scheme. *Statistics,* 47, 191-201.

Bairamov, I. G. (1997). Some distribution free properties of statistics based on record values and characterizations of the distributions through a record. *J. Applied Statist. Science*, 5, no. 1, 17-25.

Bairamov, I. G. and Ahsanullah, M. (2000). Distributional Relations between Order Statistics and the Sample itself and Characterizations of Exponential Distribution. *J. Appl. Statist. Sci*., Vol. 10(1), 1-16.

Bairamov, I. G. and Eryilmaz, S. N. (2000). Distributional properties of statistics Based on minimal spacing and record exeedance statistics. *Journal of Statistical Planning and Inference*, 90, 21-33.

Bairamov, I. G., Ahsanullah, M. and Pakes, A. G. (2005). A characterization of continuous distributions via regression on pairs of record values. Australian and New Zealand *J. Statistics*, Vol. 474, 243-247.

Bairamov, I. G., Gebizlioglu, O. L. and Kaya, M. F. (2001). Asymptotic distributions of the statistics based on order statistics, record values and invariant confidence intervals. In: *Asymptotic methods in Probability and Statistics with Applications*, Ed. By Balakrishnan, N., Ibragimov, I. A. and Nevzorov, V. B., Birkhauser, Boston, NY, 309-320.

Bairamov, M., Ahsanullah, M. and Akhundov, I. (2002). A Residual Life Function of a System Having Parallel or Series Structure. *J. of Statistical Theory & Applications*, 119-132.

Balabekyan, V. A. and Nevzorov, V. B. (1986). On number of records in a sequence of series of nonidentically distributed random variables. In: *Rings and Modules. Limit Theorems of Probability Theory* (Eds. Z. I. Borevich and V. V. Petrov) V. 1, 147-153, Leningrad, Leningrad State University (in Russian).

Balakrishnan N. and Nevzorov V. (1997). Stirling numbers and records. *Advances in Combinatorial Methods and Applications to Probability and Statistics* (N. Balakrishnan, ed.), Birkhauser, Boston, 189-200.

Balakrishnan N. and Nevzorov V. (1998). Record of records. In: *Handbook of Statistics,* v. 16 (eds.: N. Balakrishnan, C. R. Rao), Amsterdam, North Holland, 515-570.

Balakrishnan N. and Nevzorov V. (2006). Record values and record statistics- *Encyclopedia of Statistical Sciences. The Second Edition.* Wiley- Interscience. V. 10, 6995-7006.

Balakrishnan N., Nevzorov V. and Nevzorova L. (1997). Characterizations of distributions by extremes and records in Archimedean copula processes In: *Advances in the Theory and*

Practice of Statistics. A volume: In Honor of Samuel Kotz (Eds: N. L. Johnson & N. Balakrishnan) (1997), NY: Wiley, 469-478.

Balakrishnan, N. and Ahsanullah, M. (1993a). Relations for Single and Product Moments of Record Values from Exponential Distribution. *J. Appl. Statist. Sci.*, 2(1), 73-88.

Balakrishnan, N. and Ahsanullah, M. (1993b). *Relations for Single and Product Moments of Record Values from Lomax Distribution.* Sankhya, B, 56, 140-146.

Balakrishnan, N. and Ahsanullah, M. (1994a). Recurrence relations for single and product moments of record values from generalized Pareto distribution. *Commun in Statistics- Theory and Methods,* 23, 2841-2852.

Balakrishnan, N. and Ahsanullah, M. (1994b). Relations for single and product moments of record values from Lomax distribution. *Sankhya, Ser. B*, 56, 140-146.

Balakrishnan, N. and Ahsanullah, M. (1995). Relations for single and product moments of record values from exponential distribution. *J. of Applied Statistical Science*, 2, 73-87.

Balakrishnan, N. and Balasubramanian, K. (1995). A Characterization of Geometric Distribution Based on Record Values. *J. Appl. Statist. Sci.,* 2, 277-282.

Balakrishnan, N. and Stepanov, A. (2004). Two characterizations based on order statistics and records. *Journal of Statistical Planning and Inference,* 124, 273-287.

Balakrishnan, N. and Stepanov, A. (2005). A note on the number of Observations registered near an order statistic. *Journal of Statistical Planning and Inference*, 134, 1-14.

Balakrishnan, N. and Stepanov, A. (2006). On the Fisher information in record data. *Statistics and Probability Letters* 76, 537-545.

Balakrishnan, N. and Stepanov. A. (2013). Runs based on records: Their Distributional properties and an application to testing for dispersive ordering. *Methodology and Computing in Applied Probability,* 15, 583-594.

Balakrishnan, N., Ahsanullah, M. and Chan, P. S. (1992). Relations for Single and Product Moments of Record Values from Gumbel Distribution. Statist. *Probab. Letters*, 15 (3), 223-227.

Balakrishnan, N., Ahsanullah, M. and Chan, P. S. (1995). On the logistic record values and associated inference. *J. Appl. Statist. Science*, 2, no. 3, 233-248.

Balakrishnan, N., Balasubramanian, K., and Panchapakesan, S. (1997). δ-Exceedance records. *J. Appl. Statist. Science* 4(2-3), 123-132.

Balakrishnan, N., Chan, P. S. and Ahsanullah, M. (1993). Recurrence relations for moments of record values from generalized extreme value distribution. *Communications in Statistics: Theory and Methods,* 22, 1471-1482.

Balakrishnan, N., Dembinska, A. and Stepanov, A. (2008). Precedence-type tests based on record values. *Metrika,* 68, 233-255.

Balakrishnan, N., Pakes, A. and Stepanov, A. (2005). On the number and sum of Near record observations. *Advances in Applied Probability,* 37, 1-16.

Ballerini, R. (1994). A dependent F^{α}-scheme. *Statist. and Prob. Letters* 21, 21-25.

Ballerini, R. and Resnick, S. (1985). Records from improving populations. *J. Appl. Probab.* 22, 487-502.

Ballerini, R. and Resnick, S. (1987). Records in the presence of a linear trend. *Adv. Appl. Probab.*, 19, 801-828.

Barakat, H. M. (1998). Weak limit of sample extremal quotient. *Austral. & New Zealand J. Statist.*, 40, 83-93.

Barakat, H. M. and Nigm, E. M. (1996). Weak limit of sample geometric range. *J. Indian Statist. Ass.*, 34, 85-95.

Barndorff-Nielsen, O. (1964). On the limit distribution of the maximum of a random number of independent random variables. *Acta. Math* 15, 399-403.

Basak, P. (1996). Lower Record Values and Characterization of Exponential Distribution. *Calcutta Statistical Association Bulletin,* 46, 1-7.

Basak, P. and Bagchi, P. (1990). Application of Laplace approximation to record values. *Commun. Statist: Theory Methods*, 19, no.5, 1875-1888.

Basu, A. P. and Singh, B. (1998). Order statistics in exponential distribution. In: *BR2*, pp. 3-23.

Bau, J. J., Chen, H. J., and Xiong, M. (1993b). Percentage points of the studentized range test for dispersion of normal means. *J. Statist. Comput. Simul.* 44, 149-63.

Beesack, P. R. (1973). On bounds for the range of ordered variates. *Univ. Beograd, Publ. Elektrotehn, Fak, Ser. Mat. Fiz.* 412-60, 93-6.

Beesack, P. R. (1976). On bounds for the range of ordered variates II. *Aequationes Math.* 14, 293-301.

Beg, M. I. and Ahsanullah, M. (2006). On characterizating distributions by conditional expectations of functions of generalized order statistics. *J. Appl. Statist. Sci.,* Vol. 15(2), 229-244.

Beirlant, J., Teugels, J. L. and Vynckier, P. (1998). Some thoughts on extreme values. In: Accardi, L. and Heyde, C. C. (eds.), *Probability Towards 2000, Lecture Notes in Statistics*, 128, pp. 58-73. Springer, New York.

Bell, C. B. and Sarma, Y. R. K. (1980). A characterization of the exponential distribution based on order statistics. *Metrika,* 27, 263-269.

Bell, C. B., Blackwell, D., and Breiman, L. (1960). On the completeness of order statistics. *Ann. Math. Statist.* 31, 794-7.

Berred A. (1991). Record values and the estimation of the Weibull tail-coefficient. *C. R. Acad. Sci. Paris, Ser.* I 312, no. 12, 943-946.

Berred A. (1992). On record values and the exponent of a distribution with regularly varying upper tail. *J. Appl. Probab.*, 29, no. 3, 575-586.

Berred A. (1995). K-record values and the extreme value index. *J. Statist. Plann. Inference,* 45, no. 1/2, 49-64.

Berred A., Nevzorov V. and Wey S. (2005). Normalizing constants for record values in Archimedean copula processes. *J. Statist. Plann. and Infer.,* 133, 159-172.

Berred, A. (1998). An estimator of the exponent of regular variation based on k-Record values. *Statistics,* 32, 93-109.

Berred, A. (1998a). Prediction of record values. *Communications in Statistics: Theory and Methods,* 27, no. 9, 2221-2240.

Bhattacharya, P. K. (1974). Convergence of sample paths or normalized sums of induced order statistics. *Ann. Statist.* 2, 1034-9.

Bhattacharyya, G. K. (1985). The asymptotics of maximum likelihood and related estimators based on Type II censored data. *Journal of American Statistical Association,* 80, 398-404.

Bickel, P. J. (1967). Some contributions to the theory of order statistics. *Proc. 5th Berkeley Symposium,* 1, 575-91.

Bieniek, M. and Szynal, D. (2003). Characterizations of Distributions via Linearity of Regression of Generalized Order Statistics. *Metrika,* 58, 259-271.

Biondini, R. and Siddiqui, M. M. (1975). Record values in Markov sequences. *Statistical Inference and Related Topics* 2, 291-352. New York, Academic Press.

Bjerve, S. (1977). Error bounds for linear combinations of order statistics. *Ann. Statist.* 5, 357-369.

Bland, R. P., Gilbert, R. D., Kapadia, C. H., and Owen D. B. (1966). On the distributions of the range and mean range for samples from a normal distribution. *Biometrika* 53, 245-248. Correction 58, 245.

Blom, G. (1980). Extrapolation of linear estimates to larger sample sizes. *J. Amer. Statist. Ass.* 75, 912- 917.

Blom, G. (1988). Om record. *Elementa* 71, no. 2, 67-69.

Blom, G. and Holst L. (1986). Random walks of ordered elements with applications. *Amer. Statistician* 40, 271-274.

Blom, G., Thorburn, D. and Vessey, T. (1990). The Distribution of the Record Position and its Applications. *Amer. Statistician*, 44, 152-153.

Borenius, G. (1959). On the distribution of the extreme values in a sample from a normal distribution. *Skand. Aktuarietidskr*. 1958, 131-66.

Borovkov, K. and Pfeifer, D. (1995). On record indices and record times. *J. Statist. Plann. Inference*, 45, no. 1/2, 65-79.

Box, G. E. P. (1953). Non-normality and test on variances. *Biometrika* 40, 318-35.

Breiter, M. C. and Krishnaiah, P. R. (1968). Tables for the moments of gamma order statistics. *Sankhyā B* 30, 59-72.

Browne, S. and Bunge, J. (1995). Random record processes and state dependent thinning. *Stochastic Process. Appl.,* 55, 131-142.

Bruss, F. T. (1988). Invariant Record Processes and Applications to Best Choice Modeling. *Stoch. Proc. Appl.* 17, 331-338.

BuHamra, S. and Ahsanullah, M. (2013). On concomitants of bivariate Farlie-Gumbel-Morgenstern distributions. *Pakistan Journal of Statistics*, 29(9), 453-466.

Bunge, J. A. and Nagaraja, H. N. (1991). The distributions of certain record statistics from a random number of observations. *Stochastic Process. Appl.,* 38, no. 1, 167-183.

Bunge, J. A. and Nagaraja, H. N. (1992a). Dependence structure of Poisson-paced records. *J. Appl. Probab.*, 29, no. 3, 587-596.

Bunge, J. A. and Nagaraja, H. N. (1992b). Exact distribution theory for some point process record models, *Adv. In Appl. Probab.*, 24, no. 1, 20-44.

Burkschat, M. Cramer, E. and kamps, U. (2003). Dual generalized order staistics. *Metron* 61, 13-36.

Cadwell, J. H. (1954). The probability integral of range for samples from a symmetrical unimodal population. *Ann. Math. Statist.* 25, 803-806.

Carlin, B. P. and Gelfand, A. E. (1993). Parametric likelihood inference for record breaking problems. *Biometrika*, 80, 507-515.

Castano-Martinez, A., Lopez-Blazquez, F. and Salamanca-Mino, B. (2013). An Additive property of weak records from geometric distributions. *Metrika*, 76, 449-458.

Castillo, E. *Extreme Value Theory in Engineering*, 1988. Academic Inc. New York, USA.

Chan, L. K. (1967). On a characterization of distributions by expected values of extreme order statistics. *Amer. Math. Monthly*, 74, 950-951.

Chandler, K. N. (1952). The distribution and frequency of record values. *J. Roy. Statist. Soc.*, Ser. B, 14, 220-228.

Cheng, Shi-hong (1987). Records of exchangeable sequences. *Acta Math. Appl. Sinica,* 10, no. 4, 464-471.

Chow, Y. S. and Robbins, H. (1961). On sums of independent random variables with infinite moments and fair games. *Proc. Nat. Acad. Sci. USA,* 47, 330-335.

Cinlar, E. (1975). *Introduction to Stochastic Processes*. Prentice-Hall, New Jersey.

Consul, P. C. (1969). On the exact distributions of Votaw's criteria for testing compound symmetry of a covariance matrix. *Ann. Math. Stat.* 40, 836-843.

Cramer, E., Kamps, U. and Keseling, C. (2004). Characterizations via linearity of ordered random variables: A unifying approach. *Commun. Statist. Theory-Methods,* Vol. 33, No. 12, 2885-2911.

Cramer, E., Kamps, U. Relations for expectations of generalized order statistics. *J. Statis. Plann. Infer.* 89,77-89.

Cramer, H. (1936). Uber eine Eigensschaft der normalen Verteilungsfunction [About a property of the normal distribution function]. *Math. Z.*, 41, 405-414.

Crawford, G. B. (1966). Characterizations of geometric and exponential distributions. *Ann. Math. Statist,* 37, 1790-1794.

Csorgo, M. Seshadi, V. and Yalovsky, M. (1975). Applications of characterizations in the area of goodness of fit. In: G. P. Patil et al. eds. *Statististical distributions in Scientific Work*, Vol. 2, Reidel Dordrecht. 79-90.

Dallas, A .C. (1989). Some properties of record values coming from the geometric *distribution. Ann. Inst. Statist. Math.*, 41, no. 4, 661-669.

Dallas, A. C. (1981a). Record Values and the Exponential Distribution. *J. Appl. Prob.,* 18, 959-951.

Dallas, A. C. (1981b). A Characterization Using Conditional Variance. *Metrika,* 28, 151-153.

Dallas, A. C. (1982). Some results on record values from the exponential and Weibull law. *Acta Math. Acad. Sci. Hungary*, 40, no. 3-4, 307-311.

Danielak, K. and Dembinska, A. (2007a). Some characterizations of discrete distributions based on weak records. *Statistical Papers,* 48, 479-489.

Danielak, K. and Dembinska, A. (2007b). On characterizing discrete distributions via conditional expectations of weak record values, *Metrika*, 66, 129-138.

David, H. A. and Nagaraja, H. N. (2003). *Order Statistics*. Third Edition, Wiley, Hoboken, New York.

Davies, P. L.and Shabbhag, D. N. (1987). A generalization of a theorem by Deny with applications in characterization theory. *Quart, J. Math.* Oxford, 38(2), 13-34.

De Haan, I. (1070). 30,161-172. Sample extremes: an elementary introduction. *Statistics Neerlandica.*

De Haan, L. and Resnick, S. I. (1973). Almost Sure Limit Points of Record Values. *J. Appl. Prob.*, 10, 528-542.

De Haan, L. and Verkade, E. (1987). On extreme-value theory in the presence of a trend. *J. Appl. Prob.* 24, 62-76.

Deheuvels P., Nevzorov V. B. (1994b). Limit laws for k-record times. *J. Stat. Plann. Infer*, 38, 279-308.

Deheuvels, P. (1982). Spacing, record times and extremal processes. In: *Exchangeability in Probability and Statistics*. North Holland/ Elsevier, Amsterdam, 233-243.

Deheuvels, P. (1983). The Complete Characterization of the Upper and Lower Class of Record and Inter - Record Times of i.i.d. Sequence. Zeit. *Wahrscheinlichkeits theorie Verw. Geb.*, 62, 1-6.

Deheuvels, P. (1984 b). On record times associated with k-th extremes. *Proc. of the 73rd Pannonian Symp. on Math. Statist., Visegrad, Hungary, 13-18 Sept. 1982.* Budapest, 43-51.

Deheuvels, P. (1984a). The Characterization of Distributions by Order Statistics and Record Values: A Unified Approach. *J. Appl. Probability*, 21, 326-334. Correction in: *J. Appl. Probability*, 22 (1985), 997.

Deheuvels, P. (1984c). *Strong approximations of records and record times. Statistical extremes and applications*, Proc. NATO Adv. Study Inst., Reidel, Dordrecht, 491-496.

Deheuvels, P. (1988). Strong approximations of k-th records and k-th record times by Wiener processes. *Probab. Theory Rel. Fields* , 77, n. 2, 195-209.

Deheuvels, P. and Nevzorov, V .B. (1993). Records in F^{α}-scheme. I. Martingale properties. *Zapiski Nauchn. Semin. POMI* 207, 19-36 (in Russian). Translated version in *J. Math. Sci.* 81 (1996), 2368-78.

Deheuvels, P. and Nevzorov, V. B. (1999). Bootstrap for maxima and records. *Zapiski nauchnyh seminarov POMI*, v.260, 119-129 (in Russian).

Deheuvels, P. and Nevzorov, V.B. (1994a). Records in F^{α}-scheme.II. Limit theorems. *Zapiski Nauchn. Semin. POMI* 216, 42-51 (in Russian). Translated version in *J. Math. Sci.* 88 (1998), 29-35.

Deken, J .G. (1978). Record Values, Scheduled Maxima Sequences. *J. Appl. Prob.*, 15, 491-496.

Dembinska, A. (2007). Review on Characterizations of Discrete Distributions Based on Records and kth Records. *Communications in Statistics-Theory and Methods*, 37, no. 7, 1381-1387.

Dembinska, A. and Lopez-Blazquez, F. (2005). A characterization of geometric distribution through kth weak record. *Communications in Statistics-Theory and Methods*, 34, 2345-2351.

Dembinska, A. and Stepanov, A. (2006). Limit theorems for the ratio of weak records. *Statistics and Probability Letters*, 76, 1454-1464.

Dembińska, A. and Wesolowski, J. (1998). Linearity of Regression for Non-Adjacent Order Statistics. *Metrika*, 48, 215-222.

Dembińska, A. and Wesolowski, J. (2000). Linearity of Regression for Non-Adjacent Record values. *J. of Statistical Planning and Inference*, 90, 195-205.

Dembinska, A. and Wesolowski, J. (2003). Constancy of Regression for size two Record Spacings. *Pak. J. Statist.*, 19(1), 143-149.

Desu, M. M. (1971). A characterization of the exponential distribution by order statistics. *Ann. Math. Statist*, 42, 837-838.

Devroy, L. (1988). Applications of the theory of records in the study of random trees. *Acta Informatica*, 26, 123-130.

Devroy, L. (1993). Records, the maximal layer and uniform distributions in monotone sets. *Comput. Math. Appl.*, 25, no.5, 19-31.

Diersen, J. and Trenkler, G. (1985). Records tests for trend in location. *Statistics*, 28, 1-12.

Doob, J. L. (1953). *Stochastic processes.* Wiley, New York, NY.

Dufur, R. (1982). *Tests d'ajustement pour des echantillon tonques ou censures [Adjustment tests for tonics or censors samples]*. Ph.D. Thesis, Universite de Montreal.

Dufur, R., Maag, V. R. and van Eeden, C. (1984). Correcting a proof of a characterization of the exponential distribution. *J. Roy. Statist. Soc.,* B 46, 238-241.

Dunsmore, J. R. (1983). The Future Occurrence of Records. *Ann. Inst. Stat. Math.,* 35, 267-277.

Dziubdziela, W. (1990). O czasach recordowych I liczbie rekordow w ciagu zmiennych losowych [On the record times and the number of records in the course of random variables]. *Roczniki Polsk. Tow. Mat.*, Ser 2, 29, no. 1, 57-70 (in Polish).

Dziubdziela, W. and Kopocinsky, B. (1976). Limiting properties of the k-th record values. *Zastos. Mat.* 15, 187-190.

Embrechts, P. and Omey, E. (1983). On Subordinated Distributions and Random Record Processes. *Ma. P. Cam. Ph.* 93, 339-353.

Ennadifi, G. (1995). Strong approximation of the number of renewal paced record times. *J. Statist. Plann. Inference*, 45, 1/2, 113-132.

Erdös, P. and Kac, M. (1939). On the Gaussian law of errors in the theory of additive functions. *Proc. Nat. Acad. Sci. USA* 25, 206-207. (see also *Amer. Journ. Math.* 62 (1940), 738-742).

Faizan, M. and Khan, M. I. (2011). A characterization of continuous distributions through lower record statistics. *ProbStat Forum*, 4, 39-43.

Faizan, M. and Khan, M. I. Some characterizatiom results based on expected alues of the generalized order statistics. *Applied Mathematics, E-Notes*, 17, 106-113.i.

Faizan, M. and Khan,M.I. (2011). Characterizations of some probability distributions by conditional expectations of dual generalized order statistics. *Journal of Statistics Sciences*, 3, 143-150.

Faizan, M., Haque, Ziaul and Ansari, M. A. Characterizations of distributions by expected values of lower record statistics with spacing. *Journal of Modern Applied Statistical Methods*. 16(2). 310-321.

Feller, W. (1957). *An Introduction to Probability Theory and Its Applications*, Vol. I, 2nd Edition. Wiley & Sons, New York, NY.

Feller, W. (1966). *An Introduction to Probability Theory and its Applications.* Vol. II, Wiley & Sons, New York, NY.

Ferguson, T. S. (1964). A characterization of the negative exponential distribution. *Ann. Math. Statist*, 35, 1199-1207.

Ferguson, T. S. (1967). On characterizing distributions by properties of order statistics. *Sankhya, Ser. A*, 29, 265-277.

Feuerverger, A. and Hall, P. (1996). On distribution-free inference for record-value data with trend. *Annals of Statistics*, 24, 2655-2678.

Fisz, M. (1958). Characterizations of some probability distributions. *Skand. Aktirarict.*, 65-67.

Fosam, E. B., Rao, C. R. and Shanbhag, D. N. (1993). Comments on some papers involving Cauchy functional equation. *Statist. Prob. Lett.*, 17, 299-302.

Foster, F. G. and Stuart, A. (1954). Distribution Free Tests in Time Series Based on the Breaking of Records. *J. Roy. Statist. Soc., B,* 16, 1-22.

Foster, F. G. and Teichroew, D (1955). A sampling experiment on the powers of the record tests for a trend in a time series. *J. Roy. Statist. Soc.,* B17, 115-121.

Franco, M. and Ruiz, J. M. (1995). On Characterization of Continuous Distributions with Adjacent Order Statistics. *Statistics,* 26, 275-385.

Franco, M. and Ruiz, J. M. (1996). On Characterization of Continuous Distributions by Conditional Expectation of Record Values. *Sankhyā, 58, Series A*, Pt. 1, 135-141.

Franco, M. and Ruiz, J. M. (1997). On Characterizations of Distributions by Expected Values of Order Statistics and Record Values with Gap. *Metrika,* 45, 107-119.

Franco, M. and Ruiz, J. M. (2001). Characterization of Discrete distributions based on conditional expectations of Record Values. *Statistical Papers*, 42,101-110.

Fréchet, M. (1927). Sur La loi probabilite de l'écart Maximum [On The Probability Law of the Maximum Gap]. *Ann. de la oc. Polonaise de Math.*, 6, 93-116.

Freudenberg, W. and Szynal, D. (1976). Limit Laws for a Random Number of Record Values. *Bull. Acad. Polon. Sci. Ser. Math. Astr. Phys.* 24, 193-199.

Freudenberg, W. and Szynal, D. (1977). On domains of attraction of record value distributions. *Colloq. Math.*, 38, 1, 129-139.

Gajek, L. (1985). Limiting properties of difference between the successive kth record values. *Probab. and Math. Statist,* 5, n .2, 221-224.

Gajek, L. and Gather, U. (1989). Characterizations of the Exponential Distribution by Failure Rate and Moment Properties of Order Statistics. Lecture Notes in Statistics, 51. *Extreme Value Theory, Proceeding*, 114-124. J. Husler, R. D. Reiss (eds.), Springer-Verlag, Berlin, Germany.

Galambos, J. (1971). On the distribution of strongly multiplicative functions. *Bull. London Math. Soc.* 3, 307-312.

Galambos, J. (1975a). Characterizations of probability distributions by properties of order statistics II. In: G. P. Patil et al. eds., *Statistical Distributions in Scientific Work*, Vol. 3, Reidel Dordrecht, 89-101.

Galambos, J. (1975b). Characterizations of probability distributions by properties of order statistics I. In: G. P. Patil et al. eds., *Statistical Distributions in Scientific Work*, Vol. 3, Reidel Dordrecht, 71-86.

Galambos, J. (1976). A remark on the asymptotic theory of sums with random size. *Math. Proc. Cambridge Philos. Soc.* 79, 531-532.

Galambos, J. (1986). On a conjecture of Kátai concerning weakly composite numbers. *Proc. Amer. Math. Soc.* 96, 215-216.

Galambos, J. (1987). *The Asymptotic Theory of Extreme Order Statistics*. Robert E. Krieger Publishing Co. Malabar, FL.

Galambos, J. and Kátai, I. (1989). A simple proof for the continuity of infinite convolution of binary random variables. *Stat. and Probab. Lett.* 7, 369-370.

Galambos, J. and Kotz, S. (1978). Characterizations of probability distributions, *Lecture Notes in Mathematics,* Vol. 675, Springer-Verlag, New York, NY.

Galambos, J. and Kotz, S. (1983). Some characterizations of the exponential distribution via properties of geometric distribution. In: P. K. Sen, ed., *Essays in Honor of Norman L. Johnson*, North Holland, Amsterdam,159-163.

Galambos, J. and Seneta, E. (1975). Record times. *Proc. Amer. Math. Soc.* 50, 383-387.

Gather, U. (1989). On a characterization of the exponential distribution by properties of order statistics, *Statist. Prob. Lett.*, 7, 93-96.

Gather, U., Kamps, U. and Schweitzer, N. (1998). Characterizations of distributions via identically distributed functions of order statistics. In: N. Balakrishnan and C. R. Rao eds., *Handbook of Statistics,* Vol. 16, 257-290.

Gaver, D. P. (1976). Random Record Models. *J. Appl. Prob.,* 13, 538-547.

Gaver, D. P. and Jacobs, P. A. (1978). Non-Homogeneously Paced Random Records and Associated Extremal Processes. *J. Appl. Prob.,* 15, 543-551.

Glick, N. (1978). Breaking Records and Breaking Boards. *Amer. Math. Monthly,* 85(1), 2-26.

Gnedenko, B. (1943). Sur la Distribution Limite du Terme Maximum d'une Serie Aletoise [On the Distribution Limit of the Maximum Term of a Serie Aletoise]. *Ann. Math.,* 44, 423-453.

Goldburger, A. S. (1962). Best Linear Unbiased Predictors in the Generalized Linear Regression Model. *J. Amer. Statist. Assoc.,* 57, 369-375.

Goldie, C. M. (1982). Differences and quotients of record values. *Stochastic Process. Appl.,* 12, no. 2, 162.

Goldie, C. M. (1989). Records, Permutations and Greatest Convex Minorants. *Math. Proc. Camb. Phil. Soc.,* 106, 169-177.

Goldie, C. M. and Maller, R. A. (1996). A point-process approach to almost sure behavior of record values and order statistics. *Adv. in Appl. Probab.,* 28, 426-462.

Goldie, C. M. and Resnick, S. I. (1989). Records in Partially Ordered Set. *Ann. Prob.,* 17, 675-689.

Goldie, C. M. and Resnick, S. I. (1995). Many Multivariate Records. *Stoch. Proc. And Their Appl.,* 59, 185-216.

Goldie, C. M. and Rogers, L. C. G. (1984). The k - Record Processes are i.i.d. Z. Wahr. *Verw. Geb,* 67, 197-211.

Gouet, R. F., Lopez, J. and Sanz, G. (2012). Central Limit Theorem for the Number of Near-Records, *Communications in Statistics: Theory and Methods*, 41(2), 309-324.

Govindrarajulu, Z. (1966). Characterizations of the exponential and power function distributions. *Scand. Aktuarietdskr.*, 49, 132-136.

Gradshteyn, I. S. and Ryzhik, I. M. (1980). *Tables of Integrals, Series, and Products, Corrected and Enlarged Edition.* Academic Press, Inc.

Grosswald, E. and Kotz, S. (1981). An Integrated Lack of Memory Property of the Exponential Distribution. *Ann. Inst. Statist. Math.*, 33, A, 205-214.

Grosswald, E., Kotz, S. and Johnson, N. L. (1979). *Characterizations of the exponential Distribution by relevaion type equations.* University of North Corolina at Chapel Hill, Institute of Statistics Memo Series #1218.

Grudzien, Z. (1979). On distribution and moments of ith record statistic with random index. *Ann. Univ. Mariae Curie Sklodowska*, Sect A 33, 89-108.

Grudzien, Z. and Szynal, D. (1985). On the Expected Values of kth Record Values and Characterizations of Distributions. In: *Probability and Statistical Decision Theory*, Vol A. (F. Konecny, J. Mogyorodi and W. Wertz, eds.) Dordrecht –Reidel, 1195-214.

Grudzien, Z. and Szynal, D. (1996). Characterizations of distributions by order statistics and record values, A Unified Approach. *J. Appl. Prob.*, 21, 326-334..

Grudzien, Z. and Szynal, D. (1997). Characterizations of uniform and exponential distributions via moments of the kth record values randomly indexed. *Applications Mathematicae*, 24, 307-314.

Guilbaud, O. (1979). Interval estimation of the median of a general distribution. *Scand. J. Statist.* 6, 29-36.

Guilbaud, O. (1985). Statistical inference about quantile class means with simple and stratified random sampling. *Sankhyā B*, 47, 272-9.

Gulati, S. and Padgett, W. J. (1992). Kernel density estimation from record-breaking data. In: *Probability and Statistical Decision Theory,* Vol. A (Konechny F., Mogurodi J. and Wertz W., eds), 197-127. Reidel, Dordrecht.

Gulati, S. and Padgett, W. J. (1994a). Nonparametric Quantitle Estimation from Record Breaking Data. *Aust. J. of Stat.*, 36, 211-223.

Gulati, S. and Padgett, W. J. (1994b). Smooth Nonparametric Estimation of Distribution and Density Function from Record Breaking Data. Comm. In *Statist. Theory-Methods*, 23, 1259-1274.

Gulati, S. and Padgett, W. J. (1994c). Smooth Nonparametric Estimation of the Hazard and Hazard Rate Functions from Record Breaking Data. *J. Statist. Plan. Inf.*, 42, 331-341.

Gulati, S. and Padgett, W. J. (1994d). Estimation of nonlinear statistical functions from record-breaking data: a review. *Nonlinear Times and Digest*, 1, 97-112.

Gumbel, E. J. (1949). Probability tables for the range. *Biometrika* 36, 142-148.

Gumbel, E. J. (1963). Statistical forecast of droughts. *Bull. I.A.S.H.* 8, 5-23.

Gumbel, E. J. (1963). *Statistics of Extremes*. Columbia University Press, New York, NY, USA Columbia,

Gumbel, E. J. and Herbach, L. H. (1951). The exact distribution of the extremal quotient. *Ann. Math. Statist.* 22, 418-426.

Gupta, R. C. (1973). A characteristic property of the exponential distribution. *Sankhya, Ser. B*, 35, 365-366.

Gupta, R. C. (1984). Relationships between order statistic and record values and some characterization results. *J. Appl. Prob.*, 21, 425-430.

Gupta, R. C. and Ahsanullah, M. (2004). Characterization results based on the conditional expectation of a function of non-adjacent order statistic (Record Value). *Annals of Statistical Mathematics*, Vol. 56(4), 721-732.

Gupta, R. C. and Ahsanullah, M. (2004). Some characterization results based on the conditional expectation of truncated order statistics (Record Values). *Journal of Statistical Theory and Applications,* vol. 5, no. 4,391-402.

Gupta, R. C. and Kirmani, S. N. U. A. (1988). Closure and monotonicity properties of nonhomogeneous Poisson processes and record values. *Probability in Engineering and Informational Sciences,* 2, 475-484.

Gut, A. (1990). Convergence Rates for Record Times and the Associated Covering Processes. *Stoch. Processes Appl.,* 36, 135-152.

Guthree, G. L. and Holmes, P. T. (1975). On record and inter-record times for a sequence of random variables defined on a Markov chain. Adv. in *Appl. Probab.,* 7, no. 1, 195-214.

Haghighi-Talab, D. and Wright, C. (1973). On the distribution of records in a finite sequence of observations with an application to a road traffic problem. *J. Appl. Probab.,* 10, no. 3, 556-571.

Haiman, G. (1987). Almost sure asymptotic behavior of the record and record time sequences of a stationary Gaussian process. *Mathematical Statistics and Probability Theory* (M. L. Puri, P. Revesz and W. Wertz, eds.), vol. A, Reidel, Dordrecht, 105-120.

Haiman, G. and Nevzorov, V. B. (1995). Stochastic ordering of the number of records. In: *Statistical Theory and Applications: Papers in Honor of H. A. David.* (Eds.: H. N. Nagaraja, P. K. Sen and D. F. Morrison) Springer – Verlag, Berlin, 105-116.

Haiman, G., Mayeur, N., Nevzorov, V. B. and M. L. Puri M. L. (1998). Records and 2-block records of 1-dependent stationary sequences under local dependence. *Ann. Instit. Henri Poincare* v. 34, (1998), 481-503.

Hamedani, G. G. (2002). Charcterizations of univariate continuous distributions II. *Studia ScientiarumMathematicarum Hungarica*, 39, 407-424.

Hamedani, G. G. (2006). Charcterizations of univariate continuous distributions III. *Studia ScientiarumMathematicarum Hungarica*, 43, 361-385.

Hamedani, G. G. (2010). Charcterizations of continuous univariate distributions based on the truncated moments of functions of order statistics. *Studia Scientiarum Mathematicarum Hungarica*, 47,462-484.

Hamedani, G. G., Ahsanullah, M. and Sheng, R. (2008). Characterizations of certain continuous univariate distributions based on the truncated moment of the first order statistic. *Aligarh Journal of Statistics*, 28, 75-81.

Hashorva, E. and Stepanov, A. (2012). Limit theorems for the spacing of weak records. *Metrika,* 75, 163-180.

Hijab, O. and Ahsanullah, M. (2006). Weak records of geometric distribution and some characterizations. *Pakistan Journal of Statistics*, Vol. 2 no. 2, 139-146.

Hill, B. M. (1975). A simple general approach to inference about the tail of a distribution. *Ann. Statist.*, 3, 1163-1174.

Hofmann, G. (2004). Comparing Fisher information in record data and random observations. *Statistical Papers*, 45, 517-528.

Hofmann, G. and Balakrishnan, N. (2004). Fisher information in - records. *Annals of the Institute of Statistical Mathematics*, 56, 383-396.

Hofmann, G. and Nagaraja H. N. (2003). Fisher information in record data. *Metrika*, 57, 177-193.

Hofmann, G., Balakrishnan, N. and Ahmadi, J. (2005). A characterization of the factorization of the hazard function by the Fisher information in minima and upper record values. *Statistics and Probability Letters*, 72, 51-57.

Hoinkes, L. A. and Padgett, W. J. (1994). Maximum likelihood estimation from record-breaking data for the Weibull distribution. *Quality and Reliability Engineering International*, 10. 5-13.

Holmes, P. T. and Strawderman, W. (1969). A note on the waiting times between record observations. *J. Appl. Prob.*, 6, 711-714.

Huang, J. S. (1974). Characterizations of exponential distribution by order statistics. *J. Appl. Prob.,* 11, 605-609.

Huang, J. S., Arnold, B. C. and Ghosh, M. (1979). On characterizations of the uniform distribution based on identically distributed spacings. *Sankhya, Ser B*, 41, 109-115.

Huang, W. J. and Li, S. H. (1993). Characterization results based on record values. *Statistica Sinica*, 3, 83-599.

Huang, Wen-Jang and Li, Shun-Hwa (1993). Characterization Results Based on Record Values. *Statistica Sinica, 3,* 583-599.

Huang, Wen-Jang and Su, Nan-Cheng. Characterizations of Distributions Based on moments of Residual life. *Communications in Statistics. Theory and Methods.* vol., 2750-2761.

Imlahi, A. (1993). Functional laws of the iterated logarithm for records. *J. Statist. Plann. Inference*, 45, no. 1/2, 215-224.

Iwinska, M. (1985). On a characterization of the exponential distribution by order statistics. In: *Numerical Methods and Their Applications. Proc. 8th Sess Poznan Circle Zesz Nauk Ser 1.* Akad. Ekad. Poznan, 132, 51-54.

Iwinska, M. (1986). On the characterizations of the exponential distribution by record values. *Fasc. Math.*, 15, 159-164.

Iwinska, M. (1987). On the characterizations of the exponential distribution by order statistics and record values. *Fasciculi Mathematici*, 16, 101-107.

Iwinska, M. (2005). On characterization of the exponential distribution by record values with a random index. *Fasciculi Mathematici*, 36, 33-39.

Jenkinson, A. F. (1955). The Frequency Distribution of the Annual Maximum (or Minimum) Values of Meteorological Elements. *Quart. J. Meter. Soc.,* 87, 158-171.

Johnson, N. L. and Kotz, S. (1977). *Distributions in Statistics: Continuous Multivariate Distributions.* Wiley & Sons, New York, NY.

Joshi, P. C. (1978). Recurrence relation between moments of order statistics from exponential and truncated exponential distribution. *Sankhya Ser B* 39,362-371.r.

Kaigh, W. D. and Sorto, M. A. (1993). Subsampling quantile estimator majorization inequalities. *Statist. Probab. Lett.* 18, 373-379.

Kakosyan, A. V., Klebanov, L. B. and Melamed, J. A. (1984). Characterization of Distribution by the Method of Intensively Monotone Operators. *Lecture Notes in Math.* 1088, Springer Verlag, New York, N.Y.

Kaluszka, M. and Okolewski, A. (2001). An extension of the Erdös-Neveu-Rényi theorem with applications to order statistics. *Statist. Probab. Lett.* 55, 181-186.

Kaminsky, K. S. (1972). Confidence intervals for the exponential scale parameter using optimally selected order statistics. *Technometrics,* 14, 371-383.

Kaminsky, K. S. and Nelson, P. L. (1975). Best linear unbiased prediction of order statistics in location and scale families. *J. Amer. Statist. Assoc.,* 70, 145-150.

Kaminsky, K. S. and Rhodin, L. S. (1978). The prediction information in the latest failure. *J. Amer. Statist. Ass.* 73, 863-866.

Kaminsky, K. S. and Rhodin, L. S. (1985). Maximum likelihood prediction. *Ann. Inst. Statist. Math.* 37, 507-517.

Kamps, U, and Gather, U. (1997). Characteristic properties of generalized order statistics from exponential distributions. *Applications Mathematicae*, 24, 383-391.

Kamps, U. (1991). A general recurrence relation for moments of order statistics in a class of probability distributions and characterizations. *Metrika,* 38, 215-225.

Kamps, U. (1992 b). Characterizations of the exponential distributions by equality of moments. *Allg. Statist. Archiv,* 78, 122-127.

Kamps, U. (1992a). Identities for the difference of moments of successive order statistics and record values. *Metron,* 50, 179-187.

Kamps, U. (1994). Reliability properties of record values from non-identically distributed random variables. *Comm. Statist. Theory Meth.* 23, 2101-2112.

Kamps, U. (1995). *A Concept of Generalized Order Statistics.* Teubner Stuttgart.

Kamps, U. (1998a). Order Statistics. *Generalized in Encyclopedia of Statistical Sciences.* Update Vol. 3 (S. KotzmRead, C. B. and Banks, D. L. eds. Wiley & Sons, New York, NY).

Kamps, U. (1998b). Subranges of generalized order statistics from exponential distributions. *Fasciculi Mathematici*, 28, 63-70.

Kamps, U. Cramer, E. (2001). On distribution of generalized order statistics. *Statistics,* 35, 269-280.

Karlin, S. (1966). *A First Course in Stochastic Processes.* Academic Press, New York, NY.

Katzenbeisser, W. (1990). On the joint distribution of the number of upper and lower records and the number of inversions in a random sequence. *Adv. in Appl. Probab.*, 22, 957-960.

Keseling, C. (1999). Conditional Distributions of Generalized Order Statistics and some Characterizations. *Metrika*, 49, 27-40.

Khan, A. H. and Abouammoh, A. M. (2000). Characterization of distributions by conditional expectation of order statistics. *Journal of Applied Probability and Statistics*, 1,115-131. *Statistical Science,* 9, 159-167.

Khan, A. H. and Abu-Salih, M. S. (1989). Characterizations of probability distribution by conditionl expectation of order statistics. *Metron*, 47,171-181.

Khan, A. H. and Ali, M. M. (1987). Characterizations of probability distributions by higher order gap. *Communications in Statistics-Theory and Methods,* 16, 11281-1287.

Khan, A. H. and Alzaid ,A. A. (2004). Characterization of distributions through linear regression oh non-adjacent generalized order statistics. *Journal of Applied Statistical Science*, 13, 123-136.

Khan, A. H. and Imtiyaz, A. and Ahsanullah, M. (2012). Characterizations through distributional properties of dual generalized order statistics, *Egyptian Mathematical Society*, 20, 211-214.

Khan, A. H. Khan, R. U. and Yakub, M. (2006). Characterizations of continuous distributions through conditional expectation of functions of generalized order statistics. *Journal of Applied Statistical Science.*

Khan, A. H., Faizan, M. and Haque, Z. (2010). Characterizations of continuous distributions through record statistics. *Communications Korean Mathenatical Society*, 25,485-489.

Khan, A. H., Faizan, M. and Haque, Z. (2013). Characterizations of continuous distributions via conditional expectation of bon-adjacent order statistics. *Selcuk Journal of Applied Mathematics,* 14(1), 11-20. ty, 25(3),485-489.

Khan, M. I. and Faizan, M. (2012). Some characterization results based on conditional expectation of functions dual generalized order statistics.

Khan, M. J. S., Faizan, M. and Iqrar, S. (2016). Characterization of exponential distribution by spacings of records. *Aligarh Journal of statistics,* 36, 57-62.

Khan, M. J. S., Faizan, M. and Iqrar,s. (2016). Characterization of exponential distribution by spacings of order statistics. *Asian Journal of Current Engineering and Math,* 5(6),92-83; *Pakistan journal of Statistics and Operations research,* VIII (4),789-799.

Khatri, C. G. (1962). Distributions of order statistics for discrete case. *Ann. Inst. Statist. Math 14,* 167-171.

Khatri, C. G. (1965). On the distributions of certain statistics derived by the union-intersection principle for the parameters of k rectangular populations. *J. Ind. Statist. Ass.* 3, 158-164.

Khmaladze, E., Nadareishvili, M. and Nikabadze, A. (1997). Asymptotic behaviour of several repeated records. *Statistics and Probability Letters*, 35, 49-58.

Kim, J. S., Proschan, F., and Sethuraman, J. (1988). Stochastic comparisons of order statistics, with applications in reliability. *Commun. Statist. Theory Meth.* 17, 2151-2172.

Kim, S. H. (1993). Stochastic comparisons of order statistics. *J. Korean Statist. Soc.* 22, 13-25.

King, E. P. (1952). The operating characteristic of the control chart for sample means. *Ann. Math. Statist.* 23, 384-395.

Kinoshita, K. and Resnick, S. I. (1989). Multivariate records and shape. Extreme value theory (Oberwolfach, December 6-12, 1987) (J. Husler and R. D. Reiss, eds.), *Lectures Notes Statist*, v. 51, Springer-Verlag, Berlin, 222-233.

Kirmani, S. N. U. A. and Ahsanullah, M. (2019). Representation of lower records as a function of n independent random variables. (Submitted).

Kirmani, S. N. U. A. and Beg, M. I. (1984). *On Characterization of Distributions by Expected Records*. Sankhya, Ser. A, 46, no. 3, 463-465.

Klebanov, L. B. and Melamed, J. A. (1983). A Method Associated with Characterizations of the exponential distribution. *Ann. Inst. Statist. Math.*, A, 35, 105-114.

Korwar, R. M. (1984). On Characterizing Distributions for which the Second Record has a Linear Regression on the First. *Sankhya, Ser B*, 46, 108-109.

Korwar, R. M. (1990). Some Partial Ordering Results on Record Values. *Commun. Statist. Theroy-Methods*, 19(1), 299-306.

Koshar, S. C. (1990). Some partial ordering results on record values. *Communications in Statistics: Theory Methods*, 19, no. 1, 299-306.

Koshar, S. C. (1996). A note on dispersive ordering of record values. *Calcutta Statist. Association Bulletin*, 46, 63-67.

Kotb, M. S. and Ahsanullah, M. (2013). Characterizations of probability distributions via bivariate regression of generalized order statistics. *Journal of Statistical Theory and Applications*, 12(4), 321-329.

Kotz, S. (1974). Characterizations of statistical distributions: A supplement in recent surveys. *Internal. Statist. Rev.*, 42, 39-65.

Kotz, S. and Nadarajah, S. (2000). *Extreme Value Distributions, Theory and Applications*. Imperial College Press, London, U.K.

Lau, Ka-sing and Rao, C. R. Integrated Cauchy Functional Equation and Characterization of the Exponential. *Sankhya, Ser A*, 44, 72-90.

Leadbetter, M. R., Lindgreen, G. and Rootzen, H. (1983). *Extremes and Related Properties of Random Sequences and Series*, Springer-Verlag, New York, N.Y.

Lee Min-Young (2001). On a characterization of the exoponential distribution by conditional expectations of record values. *Commun. Korean Math. Soc.*, 16, 287-290.

Lee, Min-Young, Cheng, S. K and Jung, K. H. (2002). Characterizations of the exponential distribution by order statistics and conditional expectations of random variables. *Commun. Korean Math. Soc.*, 17, 39-65.

Leslie, J. R. and van Eeden, C. (1993). On a characterization of the exponential distribution on a Type 2 right censored sample. *Ann. Statist.*, 21, 1640-1647.

Li, Y. (1999). A note on the number of records near maximum. *Statist. Probab. Lett*, 43, 153-158.

Li, Y. and Pakes, A. (1998). On the number of near-records after the maximumobservation in a continuous sample. *Communications in Statistics: Theory and Methods*, 27, 673-686.

Lien, D. D., Balakrishnan, N., and Balasubramanian, K. (1992). Moments of order statistics from a non-overlapping mixture model with applications to truncated Laplace distribution. *Commun. Statist. Theory Meth.* 21, 1909-1928.

Lin, G. D. (1987). On characterizations of distributions via moments of record values. *Probab. Th. Rel. Fields*, 74, 479-483.

Lin, G. D. (1989). The product moments of order statistics with applications to characterizations of distributions. *J. Statist. Plann. Inf.* 21, 395-406.

Lin, G. D. and Huang, J. S. (1987). A note on the sequence of expectations of maxima and of record values, *Sunkhya, A* 49, no. 2, 272-273.

Lin, G. D. and Too, Y. H. (1989). Characterizations of uniform and exponential distributions. *Statist. Prob. Lett.*, 7, 357-359.

Liu, J. (1992). Precedence probabilities and their applications. *Commun. Statist. Theory Meth.* 21, 1667-1682.

Lloyd, E. H. (1952). Least squares estimation of location and scale parameters using order statistics. *Biometrika*, 39, 88-95.

Lopez-Blaquez, F. and Moreno-Rebollo, J.L. (1997). A characterization of distributions based on linear regression of order statistics and random variables. *Sankhya, Ser A*, 59, 311-323.

Maag, U., Dufour, R. and van Eeden, C. (1984). Correcting a proof of a characterization of the exponential distribution. *J. Roy. Statist. Soc.*, B, 46, 238-241.

Malinoska, I. and Szynal, (2008). D. Ob characterizatoion of certain distributions of k-th lower (upper) record values. *Appl. Math. & Computation.* 202, 338-347o.

Malov, S. V. (1997). Sequential τ-ranks. *J. Appl. Statist. Sci.*, 5, 211-224.

Mann, N. R. (1969). Optimum Estimators for Linear Functions of Location and Scale Parameters. *Ann. Math. Statist*, 40, 2149-2155.

Marsglia, G. and Tubilla, A. (1975). A Note on the Lack of Memory Property of the Exponential Distribution. Ann. Prob., 3, 352-354.

Marshall, A. W. and Olkin, I. A new method for adding a parameter to a family of distributions with applications to the exponential and Weibull families. *Biometrika* 84, 1997, 641-652.

Maswadah, M., Seham, A. M. and Ahsanullah, M. (2013). Bayesian inference on the generalized gamma distribution based on generalized order statistics. *Journal of Statistical Theory and Applications*, 12(4), 356-377.

Mellon, B. (1988). *The Olympic Record Book*. Garland Publishing, Inc. New York, NY.

Menon, M. V. and Seshardi, V. (1975). A characterization theorem useful in hypothesis testing in contributed papers. *40th session of the Inrtnal. Statist. Inst. Voorburg*, 566-590.

Mileseic, B. and eadoic, M. (2016). Some Characterizations of the Exponential Distribution based on Order Statistics. *Appl. Anal. Discrete Math*. 10,394- 407.

Mohan, N. R. and Nayak, S. S. (1982). A Characterization Based on the Equidistribution of the First Two Spacings of Record Values. *Z. Wahr. Verw. Geb*, 60, 219-221.

Nagaraja, H. N. (1977). On a characterization based on record values. *Austral. J. Statist.*, 19, 70-73.

Nagaraja, H. N. (1978). On the expected values of record values. *Austral. J. Statist.*, 20, 176-182.

Nagaraja, H. N. (1981). Some finite sample results for the selection differential. *Ann. Inst. Statist. Math*. 33, 437-448.

Nagaraja, H. N. (1982a). Some asymptotic results for the induced selection differential. *J. Appl. Probab*. 19, 253-261.

Nagaraja, H. N. (1982b). Record values and related statistics: A review. *Commun. in Statistics- Theory and Methods*, 17, 2223-2238.

Nagaraja, H. N. (1984). Asymptotic linear prediction of extreme order statistics. *Ann. Inst. Statist. Math*. 36, 2892-99.

Nagaraja, H. N. (1988a). Record Values and Related Statistics -- a Review, *Commun. Statist. Theory and Methods*, 17, 2223-2238.

Nagaraja, H. N. (1988b). Some characterizations of continuous distributions based on regression of adjacent order statistics of random variables. *Sankhya, Ser A*, 50, 70-73.

Nagaraja, H. N. (1994a). *Record occurrence in the presence of a linear trend.* Technical Report N546, Dept. Stat., Ohio State University.

Nagaraja, H. N. (1994b). Record statistics from point process models. In: *Extreme Value Theory and Applications* (Eds: Galambos J., Lechner J. and Simiu E.), 355-370, Kluwer, Dordrecht, The Netherlands.

Nagaraja, H. N. and Nevzorov, V. B. (1977). On characterizations based on recod values and order statistics. *J. Statist. Plan. Inf.*, 61, 271-284.

Nagaraja, H. N. and Nevzorov, V. B. (1996). Correlations between functions of records may be negative. *Statistics and Probability Letters* 29, 95-100.

Nagaraja, H. N. and Nevzorov, V. B. (1997). On characterizations based on record values and order statistics. *J. of Statist. Plann. and Inference* v. 63, 271-284.

Nagaraja, H. N., Sen, P. and Srivastava, R. C. (1989). Some characterizations of geometric tail distributions based on record values. *Statistical Papers*, 30, 147-155.

Nayak, S. S. (1981). Characterizations based on record values. *J. Indian Statist. Assn.*, 19, 123-127.

Nayak, S. S. (1984). Almost sure limit points of and the number of boundary crossings related to SLLN and LIL for record times and the number of record values. *Stochastic Process. Appl.*, 17, no. 1, 167-176.

Nayak, S. S. (1985). Record values for and partial maxima of a dependent sequence. *J. Indian Statist. Assn.*, 23, 109-125.

Nayak, S. S. (1989). On the tail behaviour of record values. *Sankhya A*, 51, no. 3, 390-401.

Nayak, S. S. and Inginshetty, S. (1995). On record values. *J. of Indian Society for Probability and Statistics*, 2, 43-55.

Nayak, S. S. and Wali, K. S. (1992). On the number of boundary crossings related to LIL and SLLN for record values and partial maxima of i.i.d. sequences and extremes of uniform spacings. *Stochastic Process. Appl.*, 43, no. 2, 317-329.

Neuts, M. F. (1967). Waiting times between record observations. *J. Appl. Prob.*, 4, 206-208.

Nevzorov, V. B. (1981). Limit theorems for order statistics and record values. *Abstracts of the Third Vilnius Conference on Probability Theory and Mathematical Statistics*, v. 2, 86-87.

Nevzorov, V. B. (1984 b). Record times in the case of nonidentically distributed random variables. *Theory Probab. and Appl.*, v. 29, 808-809.

Nevzorov, V. B. (1984a). Representations of order statistics, based on exponential variables with different scaling parameters. *Zap.*

Nauchn. Sem. Leningrad 136, 162-164. English translation (1986). *J. Soviet Math.* 33, 797-8.

Nevzorov, V. B. (1985). Record and interrecord times for sequences of non-identically distributed random variables. *Zapiski Nauchn. Semin. LOMI* 142, 109-118 (in Russian). Translated version in *J. Soviet. Math.* 36 (1987), 510-516.

Nevzorov, V. B. (1986 b). On k-th record times and their generalizations. *Zapiski nauchnyh seminarov LOMI*, v. 153, 115-121 (in Russian). English version: *J. Soviet. Math.*, v. 44 (1989), 510-515.

Nevzorov, V. B. (1986 c). Record times and their generalizations. *Theory Probab. and Appl.*, v. 31, 629-630.

Nevzorov, V. B. (1986a). Two characterizations using records. *Stability Problems for Stochastic Models* (V. V. Kalashnikov, B. Penkov, and V. M. Zolotarev, Eds.), *Lecture Notes in Math* 1233, 79-85. Berlin: Springer Verlag.

Nevzorov, V. B. (1987). Moments of some random variables connected with records. *Vestnik of the Leningrad Univ.* 8, 33-37 (in Russian).

Nevzorov, V. B. (1988). Centering and normalizing constants for extrema and records. *Zapiski nauchnyh seminarov LOMI*, v. 166, 103-111 (in Russian). English version: *J. Soviet. Math.*, v. 52 (1990), 2830-2833.

Nevzorov, V. B. (1988). Records. *Theo. Prob. Appl.*, 32, 201-228.

Nevzorov, V. B. (1989). Martingale methods of investigation of records. In: *Statistics and Control Random Processes*. Moscow State University, 156-160 (in Russian).

Nevzorov, V. B. (1990). Generating functions for k^{th} record values- a martingale approach. *Zapiski Nauchn. Semin. LOMI* 184, 208-214 (in Russian). Translated version in *J. Math. Sci.* 68 (1994),545-550.

Nevzorov, V. B. (1992). A characterization of exponential distributions by correlation between records. *Mathematical Methods of Statistics*, 1, 49-54.

Nevzorov, V. B. (1993a). Characterizations of certain non-stationary sequences by properties of maxima and records. *Rings and Modules.*

Limit Theorems of Probability Theory (Eds., Z. I. Borevich and V. V. Petrov), v. 3, 188-197, St.-Petersburg, St. Petersburg State University (in Russian).

Nevzorov, V. B. (1993b). Characterizations of some nonstationary sequences by properties of maxima and records. Rings and Modules. *Limit Theorems of Probability Theory*, v.3, 188-197 (in Russian).

Nevzorov, V. B. (1995). Asymptotic distributions of records in nonstationary schemes. *J. Stat. Plann. Infer.*, 45, 261-273.

Nevzorov, V. B. (1997). A limit relation between order statistics and records. *Zapiski nauchnyh seminarov LOMI*, v. 244, 218-226 (in Russian). English version: in J. Math. Sci.

Nevzorov, V. B. (2000). *Records. Mathematical Theory.* Moscow: Phasis (in Russian), 244 p.

Nevzorov, V. B. (2001). *Records: Mathematical Theory. Translation of Mathematical Monographs,* Volume 194. American Mathematical Society. Providence, RI, 164p.

Nevzorov, V. B. (2004). Record models for sport data. *Longevity, ageing and degradation models*, v. 1 (Eds: V. Antonov, C. Huber, M. Nikulin), S- Petersburg, 198-200.

Nevzorov, V. B. (2012). On the average number of records in sequences of nonidentically distributed random variables. *Vestnik of SPb State University*, v. 45, n. 4, 164-167 (in Russian).

Nevzorov, V. B. and Balakrishnan, N. (1998). A record of records. In: *Handbook of Statistics, Eds.* N. Balakrishnan and C. R. Rao, Elsevier Science, Amsterdam, pp. 515-570.

Nevzorov, V. B. and Rannen M. (1992). On record moments in sequences of nonidentically distributed discrete random variables. *Zapiski nauchnyh seminarov LOMI*, v. 194 (1992), 124-133 (in Russian). English version: in J. Math. Sci.

Nevzorov, V. B. and Saghatelyan, V. (2009). On one new model of records. *Proceedings of the Sixth St. Petersburg Workshop on Simulation* 2: 981-984.

Nevzorov, V. B. and Stepanov, A. V. (1988). Records: martingale approach to finding of moments. *Rings and Modules. Limit Theorems of ProbabilityTheory* (Eds. Z. I. Borevich and V. V. Petrov), v. 2, 171-181, St.-Petersburg, St. Petersburg State University (in Russian).

Nevzorov, V. B. and Stepanov, A. V. (2014). Records with confirmations. *Statistics and Probability Letters*, v. 95, 39-47.

Nevzorov, V. B. and Tovmasjan S. A. (2014). On the maximal value of the average number of records. *Vestnik of SPb State University*, v. 1 (59), n.2, 196-200 (in Russian).

Nevzorov, V.B. (2013). Record values with restrictions. *Vestnik of SPb State University*, v. 46, n. 3, 70-74 (in Russian).

Nevzorova, L. N. and Nevzorov, V. B. (1999). Ordered random variables. *Acta Appl. Math.*, 58, no. 1-3, 217-219.

Nevzorova, L. N., Nevzorov, V. B. and Balakrishnan, N. (1997). Characterizations of distributions by extreme and records in Archimedian copula processes. In: *Advances in the Theory and Pratice of Statistics- A volume in Honor of Samuel Kotz* (eds. N. L. Johnson and N. Balakrishnan), 469-478. John Wiley and Sons, New York, NY.

Newell, G. F. (1963). Distribution for the smallest distance between any pair of *k*th nearest neighbor random points on a line. In: *Proceedings of Symposium on Time Series Analysis* (Brown University), pp. 89-103. Wiley, New York.

Neyman, J. and Pearson, E. S. (1928). On the use and interpretation of certain test criteria for purposes of statistical inference, I. *Biometrika* 20A, 175-240.

Nigm, E. M. (1998). On the conditions for convergence of the quasi-ranges and random quasi-ranges to the same distribution. *Amer. J. Math. Mgmt. Sci.* 18, 259-76.

Noor, Z. E. and Athar, H. (2014). Characterizations of probability distributions by conditional expectations of record statistics. *Journal of Egyptian mathematical Society*. 22, 275-279.

Olkin, I. and Stephens, M. A. (1993). On making the shortlist for the selection of candidates. *Intern. Statist. Rev.* 61, 477-486.

Oncei, S. Y., Ahsanullah, M., Gebizlioglu, O. I. and Aliev, F. A. (2001). Characterization of geometric distribution through normal and weak record values. *Stochastc Modelling and Applications*, 4(1), 53-58.

Oncel,S.Y. Ahsanullah, M., F. A. Aliev and F. Aygun (2005).Switching record and order statistics via random contractions. *Statistics and Probability Letters*, 73,207-217.

Owen, D. B. (1962). *Handbook of Statistical Tables.* Addison-Wesley, Reading, MA.

Owen, D. B. and Steck, G. P. (1962). Moments of order statistics from the equicorrelated multivariate normal distribution. *Ann. Math. Statist.* 33, 1286-1291.

Pakes, A. and Steutel, F. W. (1997). On the number of records near the maximum. *Austral. J. Statist.*, 39, 179-193.

Pawlas, P. and Szynal, D. (1999). Recurrence Relations for Single and Product Moments of K-th Record Values From Pareto, Generalized Pareo and Burr Distributions. *Commun. Statist. Theory-Methods,* 28(7), 1699-1701.

Pfeifer D. (1991). Some remarks on Nevzorov's record model, *Adv. Appl. Probab.*, 23, 823-834.

Pfeifer D. and Zhang, Y. C. (1989). A survey on strong approximation techniques in connection with. *Extreme value theory* (Oberwolfach, December 6-12, 1987) (J. Husler and R. D. Reiss, eds.), *Lectures Notes Statist.*, v.51, Springer-Verlag, Berlin, 50-58.

Pfeifer, D. (1981). Asymptotic Expansions for the Mean and Variance of Logarithmic Inter-Record Times. *Meth. Operat. Res.*, 39, 113-121.

Pfeifer, D. (1982). Characterizations of exponential distributions by independent non-stationary record increments. *J. Appl. Prob.*, 19, 127-135 (Corrections 19, p. 906).

Pfeifer, D. (1984a). A note on moments of certain record statistics. *Z. Wahrsch. verw. Gebiete, Ser. B*, 66, 293-296.

Pfeifer, D. (1984b). Limit laws for inter-record times from non-homogeneous record values. *J. Organizat. Behav. and Statist.* 1, 69-74.

Pfeifer, D. (1985). On a relationship between record values and Ross' model of algorithm efficiency. *Adv. in Appl. Probab.*, 27, no .2, 470-471.

Pfeifer, D. (1986). Extremal processes, record times and strong approximation. *Publ. Inst. Statist. Univ. Paris*, 31, no. 2-3, 47-65.

Pfeifer, D. (1987). On a joint strong approximation theorem for record and inter-record times. *Probab. Theory Rel. Fields*, 75, 213-221.

Pfeifer, D. (1989). Extremal processes, secretary problems and the 1/e law. *J. of Applied Probability*, 8, 745-756.

Prescott, P. (1970). Estimation of the standard deviation of a normal population from doubly censored samples using normal scores. *Biometrika* 57, 409-419.

Prescott, P. (1971). Use of a simple range-type estimator of σ in tests of hypotheses. *Biometrika* 58, 333-340.

Pudeg, A. (1990). Charakterisierung von wahrseinlichkeitsver-teilungen durch ver teilungseigenschaften der ordnungesstatistiken und rekorde [Characterization of truth distributions by distribution characteristics of the regulation statistics and records]. Dissertation Aachen University of Technology.

Puri, M. L. and Ruymgaart, F. H. (1993). Asymptotic behavior of *L*-statistics for a large class of time series. *Ann. Inst. Statist. Math.* 45, 687-701.

Puri, P. S. and Rubin, H. (1970). A characterization based on the absolute difference of two i.i.d. random variables. *Ann. Math. Statist.*, 41, 2113-2122.

Pyke, R. (1965). Spacings. *J. Roy. Statist. Soc. B* 27, 395-436. Discussion: 437-49.

Rahimov, I. (1995). Record values of a family of branching processes. *IMA Volumes in Mathematics and Its Applications*, v. 84, Springer - Verlag, Berlin, 285-295.

Rahimov, I. and Ahsanullah, M. (2001). M. Records Generated by the total Progeny of Branching Processes. *Far East J. Theo. Statist,* 5(10), 81-84.

Rahimov, I. and Ahsanullah, M. (2003). Records Related to Sequence of Branching Stochastic Process. *Pak. J. Statist.,* 19, 73-98.

Ramachandran, B. (1979). On the strong memoryless property of the exponential and geometric laws. *Sankhya, Ser A,* 41, 244-251.

Ramachandran, B. and Lau, R. S. (1991). *Functional Equations in Probability Theory.* Academic Press, Boston, MA.

Rannen, M. M. (1991). Records in sequences of series of nonidentically distributed random variables. *Vestnik of the Leningrad State University* 24, 79-83.

Rao, C. R. and Shanbhag, D. N. (1986). Recent results on characterizations of probability distributions. A unified approach through extensions of Deny's theorem. *Adv. Appl. Prob.,* 18, 660-678.

Rao, C. R. and Shanbhag, D. N. (1994). *Choquet Deny Type Functional Equations with Applications to stochastic models.* Wiley & Sons, Chichestere.

Rao, C. R. and Shanbhag, D. N. (1995a). Characterizations of the exponential distribution by equality of moments. *Allg. Statist. Archiv,* 76, 122-127.

Rao, C. R. and Shanbhag, D. N. (1995b). *A conjecture of Dufur on a characterization of the exponential distributions.* Center for Multivariate Analysis. Penn State University Technical Report, 95-105.

Rao, C. R. and Shanbhag, D. N. (1998). Recent approaches for characterizations based on order statistics and record values. *Handbook of Statistics,* (N. Balakrishnan and C. R. Rao, eds.) Vol. 10, 231-257.

Rao. C. R. (1983). An extension of Deny's theorem and its application to characterizations of probability distributions. In: P. J. Bickel et al.

eds., *A festschrift for Erich L. Lehman,* Wordsworth, Belmont, 348-366.

Raqab, M, Z. Exponential distribution records: different methods of prediction. In: *Recent Developments in Ordered Random Variables.* Nova Science Publishers Inc. 2006, 239-261-195. Edited by M. Ahsanullah and M. Z. Raqab.

Raqab, M. Z. (1997). Bounds based on greatest convex minorants for moments of record values. *Statist. Prob. Lett.,* 36, 35-41.

Raqab, M. Z. (2002). Characterizations of distributions based on conditional expectations of record values. *Statistics and Decisions,* 20, 309-319.

Raqab, M. Z. and Ahsanullah, M. (2000). Relations for marginal and joint moment generating functions of record values from power function distribution. *J. Appl. Statist. Sci.,* Vol. 10(1), 27-36.

Raqab, M. Z. and Ahsanullah, M. (2001). Estimation of the location and scale parameters of generalized exponential distribution based on order statistics. *Journal of Statistical Computing and Simulation,* Vol. 69(2), 109-124.

Raqab, M. Z. and Ahsanullah, M. (2003). On moment generating functions of records from extreme value distributions. *Pak. J. Statist.,* 19(1), 1-13.

Raqab, M. Z. and Amin, W. A. (1997). A note on reliability properties of *k* record statistics. *Metrika,* 46, 245-251.

Reidel, M. and Rossberg, H. J. (1994). Characterizations of the exponential distribution function by properties of the difference $X_{k+s,n} - X_{k,n}$ of order statistics. *Metrika,* 41, 1-19.

Reiss, R. D. (1989). *Approximate Distributions of Order Statistics.* Springer-Verlag, New York, NY.

Rényi, A. (1962). Theorie des elements saillants d'une suit d'observations. Colloquium on Combinatorial Methods in Probability Theory, Math. *Inst., Aarhus University, Aarhus, Denmark,* August 1-10, 104-115. See also: *Selected Papers of Alfred Rényi,* Vol. 3 (1976), Akademiai Kiado, Budapest, 50-65.

Resnick, S. (1987). *Extreme Values, Regular Variation and Point Processes*. Springer-Verlag, New York, NY.

Resnick, S. I. (1973a). Limit laws for record values, *Stoch. Proc. Appl.*, 1, 67-82.

Resnick, S. I. (1973b). Record values and maxima. *Ann. Probab.*, 1, 650-662.

Resnick, S. I. (1973c). Extremal processes and record value times. *J. Appl. Probab.*, 10, no. 4, 864-868.

Roberts, E. M. (1979). Review of statistics of extreme values with application to air quality data. Part II: Application. *J. Air Polution Control Assoc.*, 29, 733-740.

Rossberg, H. J. (1972). Characterization of the exponential and Pareto distributions by means of some properties of the distributions which the differences and quotients of order statistics subject to. *Math Operationsforsch Statist.*, 3, 207-216.

Roy, D. (1990). Characterization through record values. *J. Indian. Statist. Assoc.*, 28, 99-103.

Saan, J. and Pandey,A. (200). Recurrence Relations for single and Product Moments of Generalized order statistics from linear Exponential Distribution. *Journal of Applied Saisical Science*, vol, 13, no. 4, 323-333.

Sagateljan, V. K. (2008). On one new model of record values. *Vestnik of St. Petersburg University*, Ser. 1, n. 3: 144-147 (in Russian).

Salah, M. M., M. Z. Raqab and Ahsanullah, M. (2009). *Journal of Applied Statistical Science*, vol. 17, no. 1. pp. 81-91.

Salamiego, F. J. and Whitaker, L.D. (1986). On estimating population characteristics from record breaking observations. I. Parametric Results. *Naval Res. Log. Quart*, 33, no.3, 531-543.

Samaniego, F. G. and Whitaker, L. R. (1988). On estimating population characteristics from record-breaking observations, II. Nonparametric results. *Naval Res. Logist. Quart*, 35, no. 2, 221-236.

Sarhan, A. E. and Greenberg, B. G. (1959). Estimation of location and scale parameters for the rectangular population from censored samples. *J. Roy. Statist. Soc. B* 21, 356-63.

Sarhan, A. E. and Greenberg, B. G. (eds.) (1962). *Contributions to Order Statistics.* Wiley, New York.

Sathe, Y. S. (1988). On the correlation coefficient between the first and the rth smallest order statistics based on n independent exponential random variables. *Commun. Statist. Theory Meth.* 17, 3295-9.

Sathe, Y. S. and Bendre, S. M. (1991). Log-concavity of probability of occurrence of at least r independent events. *Statist. Probab. Lett.* 11, 63-64.

Saw, J. G. (1959). Estimation of the normal population parameters given a singly censored sample. *Biometrika* 46, 150-159.

Sen, P. K. (1959). On the moments of the sample quantiles. *Calcutta Statist. Assoc. Bull.*, 9, 1-19.

Sen, P. K. and Salama, I. A. (eds.). (1992*). Order Statistics and Nonparametric Theory and Applications*. Elsevier, Amsterdam.

Seshardi, V., Csorgo, N. M. and Stephens, M. A. (1969). Tests for the exponential distribution using Kolmogrov-type statistics. *J. Roy Statist. Soc.*, B, 31, 499-509.

Shah, B. K. (1970). Note on moments of a logistic order statistics. *Ann. Math. Statist*, 41, 2151-2152.

Shah, S. M. and Jani, P. N. (1988). UMVUE of reliability and its variance in two parameter exponential life testing model. *Calcutta. Statist. Assoc. Bull.*, 37, 209-214.

Shakil, M. and Ahsanullah, (2011). M. Record values of the ratio of Rayleigh random variables. Pakistan Journal of Statistics, 27(3),307-325.

Shimizu, R. (1979). A characterization of exponential distribution involving absolute difference of i.i.d. random variables. *Proc. Amer. Math. Soc.*, 121, 237-243.

Shorrock, R. W. (1972a). A limit theorem for inter-record times. *J. Appl. Prob.,* 9, 219-223.

Shorrock, R. W. (1972b) On record values and record times. *J. Appl. Prob.*, 9, 316-326.

Shorrock, R. W. (1973). Record values and inter-record times. *J. Appl. Prob.*, 10, 543-555.

Sibuya, M. and Nishimura, K. (1997). Prediction of record-breakings. *Statistica Sinica,* 7, 893-906.

Siddiqui, M. M. and Biondini, R. W. (1975). The joint distribution of record values and inter-record times. *Ann. Prob.*, 3, 1012-1013.

Singpurwala, N. D. (1972). Extreme values from a lognormal law with applications to air population problems. *Technometrics*, 14, 703-711.

Smith, R. L. (1986). Extreme value theory based on the *r* largest annual events. *J. Hydrology*, 86, 27-43.

Smith, R. L. (1988). Forecasting records by maximum likelihood. *J. Amer. Stat. Assoc.*, 83, 331-338.

Smith, R. L. and Miller, J. E. (1986). A non-Gaussian state space model and application to prediction of records. *J. Roy. Statist. Soc., Ser B*, 48, 79-88.

Smith, R. L. and Weissman, I. (1987). Large Deviations of Tail Estimators based on the Pareto Approximations. *J. Appl. Prob.*, 24, 619-630.

Srivastava, R. C. (1978). Two Characterizations of the Geometric Distribution by Record Values. *Sankhya, Ser B*, 40, 276-278.

Srivastava, R. C. (1981a). Some characterizations of the geometric distribution. In: *Statistical Distributions in Scientific Work* (C. Tallie, G. P. Patil and A. Baldessari, eds.), v. 4, Reidel, Dordrecht, 349-356.

Srivastava, R. C. (1981b). Some characterizations of the exponential distribution based on record values. In: *Statistical Distributions in Scientific Work* (C. Tallie, G. P. Patil and A. Baldessari, eds.), v. 4, Reidel, Dordrecht, 411-416.

Stam, A. I. (1985). Independent Poisson processes generated by record values and inter-record times. *Stochastic Process. Appl.*, 19, no. 2, 315-325.

Stepanov, A. (1999). The second record time in the case of arbitrary distribution. Istatistik. *Journal of the Turkish Statistical Association*, 2 (2), 65-70.

Stepanov, A. (2001). Records when the last point of increase is an atom. *Journal of Applied Statistical Science*, 10, no. 2, 161-167.

Stepanov, A. (2003). Conditional moments of record times. *Statistical Papers,* 44, no. 1,131-140.

Stepanov, A. (2004). Random intervals based on record values. *Journal of Statistic Planning and Inference*, 118, 103-113.

Stepanov, A. (2006). The number of records within a random interval of the current record value. *Statistical Papers*, 48, 63-79.

Stepanov, A. V. (1987). On logarithmic moments for inter-record times. *Teor. Veroyatnost. i Primenen.*, 32, no. 4, 774-776. Engl. Transl. in *Theory Probab. Appl.,* 32, no. 4, 708-710.

Stepanov, A. V. (1992). Limit Theorems for Weak Records. *Theory of Probab. and Appl.*, 37, 579-574 (English Translation).

Stepanov, A. V., Balakrishnan, N. and Hofmann, G. (2003). Exact distribution and Fisher information of weak record values. *Statistics and Probability Letters*, 64,69-81.

Strawderman, W. E. and Holmes, P. T. (1970). On the law of the iterated logarithm for inter-record times. *J. Appl. Prob.* 7, 432-439.

Stuart, A. (1957). The efficiency of the records test for trend in normal regression. *J. Roy. Statist. Soc., Ser. B*, 19, no. 1, 149-153.

Sukhatme, P. V. (1937). Tests of significance of the χ^2-population with two degrees of freedom. *Ann. Eugenics*, 8, 52-56.

Tallie, C. (1981). A Note on Srivastava's Characterization of the Exponential Distribution Based on Record Values. In: C. Tallie, G. P. Patil and B. Boldessari eds., *Statistical Distributions in Scientific Work.* Reidel-Dordrecht., 4, 417-418.

Tanis, E. A. (1964). Linear forms in the order statistics from an exponential distribution. *Ann. Math. Statist*, 35 ,270-276.

Tata, M. N. (1969). On outstanding values in a sequence of random variables. *Z. Wahrscheinlichkeitstheorie Verw. Geb.*,12, 9-20.

Teichroew, D. (1956). Tables of expected values of order statistics and products of order statistics for samples of size twenty and less from the normal distribution. *Ann. Math. Stat.*, 27, 410-426.

Teugels, J. L. (1984). On successive record values in a sequence of independent and identically distributed random variables. *Statistical Extremes and Applications.* (Tiago de Oliveira, ed.). Reidel-Dordrecht, 639-650.

The Guiness Book of Records (1955, etc). Guiness Books, London.

Tiago de Oliveira, J. (1968). Extremal distributions. *Rev. Fac. Cienc. Univ. Lisboa*, A, 7, 215-227.

Tietjen, G. L., Kahaner, D. K. and Beckman, R. J. (1977). Variances and covariances of the normal order statistics for sample sizes 2 to 50. *Selected Tables in Mathematical Statistics*, 5, 1-73.

Tippett, L. H. C. (1925). On the Extreme Individuals and the Range of Samples Taken from a Normal Population. *Biometrika*, 17, 364- 387.

Tryfos, P. and Blackmore, R. (1985). Forecasting records. *J. Amer. Statist. Assoc,* 80, no. 385, 46-50.

Tsokos, Chris (2011). *Probability for Engineering, Mathematics and Sciences.* Dunbury Pres,

Vervaat, W. (1973). Limit Theorems for Records from Discrete Distributions. *Stoch. Proc. Appl.*, 1, 317-334.

Weissman, I. (1978). Estimations of Parameters and Large Quantiles Based on the k Largest Observations. *J. Amer. Statist. Assoc.*, 73, 812-815.

Weissman, I. (1995). Records from a power model of independent observations. *J. of Applied Probability*, 32, 982-990.

Wesolowski, J, and Ahsanullah, M. (2000). Linearity of Regresion for Non-adjacent weak Records. *Statistica Sinica*, 11, 30-52.

Westcott, M. (1977a). A note on record times. *J. Appl. Prob.* 14, 637-639.

Westcott, M. (1977b). The Random Record Model. *Proc. Roy. Soc. Lon., A*, 356,529-547.

Westcott, M. (1979). On the Tail Behavior of Record Time Distributions in a Random Record Process. *Ann. Prob.*, 7, 868-237.

Williams, D. (1973). On Renyi''s record problem and Engel's series. *Bull. London Math.* Soc., 5, 235-237.

Williams, D. (1973). On Renyi's record problem and Engel's series. *Bulletin of the London Math.* Society, 5, 235-237.

Witte, H. J. (1988). Some characterizations of distributions based on the integrated Cauchy functional equations. *Sankhya, Ser A*, 50, 59-63.

Witte, H. J. (1990). Characterizations of distributions of exponential or geometric type integrated lack of memory property and record values. *Comp. Statist. Dat. Anal.*, 10, 283-288.

Witte, H. J. (1993). Some characterizations of exponential or geometric distributions in a non-stationary record values models. *J. Appl. Prob.*, 30, 373-381.

Wu, J. (2001). Characterization of Generalized Mixtures of Geometric and Exponential Distributions Based on Upper Record Values. *Statistical Papers*, 42, 123-133.

Wu, J. W. (2004). On characterizing distributions by conditional expectations of functions of non-adjacent record values. *Journal of Applied Statistical Science*, vol. 13,no. 2,137-145.

Xu, J. L. and Yang, G. L. (1995). A note on a characterization of the exponential distribution based on type II censored samples. *Ann. Statist.*, 23, 769-773.

Yakimiv, A. L. (1995). Asymptotics of the kth record values. *Teor. Verojatn. I Primenen*, 40, no. 4, 925-928 (in Russian). English translation in *Theory Probab. Appl.*, 40, no. 4, 794-797.

Yanev, G. P. and Ahsanullah, M. (2009). On characterizations based on regression of linear combinations f record values. *Sankhya* 71 part 1, 100-121.

Yanev, G. P. and Chakraborty, S. (2013). Characterizations of exponential distribution based on a sample of size 3. *Pliska Stud. Math. Bulgar.* 23,101-108.

Yanev, G. P., Ahsanullah, M. and Beg, M. I. (2007). Characterizations of probability distributions via bivariate regression of record values. *Metrika*, 68, 51-64.

Yang, M. C. K. (1975). On the distribution of inter-record times in an increasing population. *J. Appl. Prob.*, 12, 148-154.

Yanushkevichius, R. (1993). Stability of characterization by record property. *Stability Problems for Stochastic Models* (Suzdal, 1991). *Lecture Notes in Mathematics*, v. 1546, Springer-Verlag, Berlin, 189-196.

Zahle, U. (1989). Self-similar random measures, their carrying dimension and application to records. *Extreme Value Theory* (Oberwolfach, December 6-12, 1987) (J. Husler and R. D. Reiss, eds.), *Lectures Notes Statist.*, v. 51, Springer-Verlag, Berlin, 59-68.

Zijstra, M. (1983). On characterizations of the geometric distributions by distributional properties. *J. Appl. Prob.*, 20, 843-850.

ABOUT THE AUTHOR

Dr. Mohammad Ahsanullah

Dr. M. Ahsanullah earned his PhD from North Carolina State University, Raleigh, North Carolina. He is a Fellow of American Statistical Association and Royal Statistical Society. He is an elected member of International Statistical Institute. He is editor-in-Chief of Journal of Statistical Theory and Applications and Book Series Editor of Applied Statistical Science of Nova Science Publishers. He has taught in several universities in USA and aboard. He has authored and co-authored of more the forty books and more than 350 research articles in reputable statistical journals. His research areas are Record Values, Order Statistics, characterizations of Distributions and Statistical Inferences etc.

INDEX

S

T

U

V

The Future of Post-Human Probability: Towards a New Theory of Objectivity and Subjectivity

Author: Peter Baofu

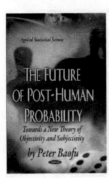

Series: Applied Statistical Science

Book Description: This book offers an alternative way to understand the future of probability, especially in the dialectic context of objectivity and subjectivity—while learning from different approaches in the literature but without favoring any one of them (nor integrating them, since they are not necessarily compatible with each other).

Hardcover ISBN: 978-1-62948-671-0
Retail Price: $325

Sequencing and Scheduling with Inaccurate Data

Editors: Yuri N. Sotskov and Frank Werner

Series: Applied Statistical Science

Book Description: During the last decades, several approaches have been developed for sequencing and scheduling with inaccurate data, depending on whether the data is given as random numbers, fuzzy numbers or whether it is uncertain, i.e., it can take values from a given interval. This book considers the four major approaches for dealing with such problems: a stochastic approach, a fuzzy approach, a robust approach and a stability approach.

Hardcover ISBN: 978-1-62948-677-2
Retail Price: $255

APPLIED STATISTICAL THEORY AND APPLICATIONS

EDITOR: Mohammad Ahsanullah

SERIES: Theoretical and Applied Mathematics

BOOK DESCRIPTION: This book makes a significant contribution to the advancement of statistical science. It contains research in many statistical designs, compares many statistical models, and includes a theory that is oriented to real life problems.

HARDCOVER ISBN: 978-1-63321-858-1
RETAIL PRICE: $230

RESEARCH IN APPLIED STATISTICAL SCIENCE

EDITOR: Mohammad Ahsanullah

SERIES: Theoretical and Applied Mathematics

BOOK DESCRIPTION: This book makes a significant contribution to the advancement of statistical science. It contains research in many statistical designs, compares many statistical models, and includes theories that are oriented to real life problems.

HARDCOVER ISBN: 978-1-63321-818-5
RETAIL PRICE: $205